Management of
Engineering Projects

Management of Engineering Projects

Victor G. Hajek

United States Naval Training Equipment Center
Orlando, Florida

McGraw-Hill Book Company

New York   St. Louis   San Francisco   Auckland   Bogotá   Düsseldorf
Johannesburg   London   Madrid   Mexico   Montreal   New Delhi   Panama
Paris   São Paulo   Singapore   Sydney   Tokyo   Toronto

Library of Congress Cataloging in Publication Data

Hajek, Victor G
    Management of engineering projects.

    First ed. published in 1965 under title:
Project engineering.
    Includes index.
    1. Engineering—Management.   2. Industrial
project management.   I. Title.
TA190.H35   1977   658.4'04   76-55763
ISBN 0-07-025534-2

1 2 3 4 5 6 7 8 9 0   K P K P   7 8 6 5 4 3 2 1 0 9 8 7

The editors for this book were Tyler G. Hicks and
Joan Zseleczky, the designer was Naomi Auerbach,
and the production supervisor was Frank P. Bel-
lantoni. It was set in Baskerville by Bi-Comp, Incorporated. Printed and bound by The Kingsport
Press.

To the project engineers, educators, secretaries, administrators, administrative assistants, Officers, and other personnel of the U.S. Naval Training Equipment Center.

Contents

Preface, xv

1 Role of the Project Engineer 1

1.1 Objectives and Disciplines, 1
1.2 Authority of Project Engineers, 4
1.3 Project Engineers' Responsibility to Management, 4
1.4 Summary, 5

2 The Procurement Project Initiated 6

2.1 Request for Proposal, 6
2.2 Participation Decision, 7
2.3 Technical Considerations, 8
2.4 Performance Specifications, 9
2.5 Analysis of Specification Requirements, 11
2.6 Technical Approach, 11
2.7 Summary, 16

3 Proposed System Design Considerations 18

3.1 Equipment Systems and Elements, 18
3.2 Tiers of Design Detail, 19
3.3 Preliminary Block Diagrams, 20
3.4 Factors Affecting System Approach, 21
3.5 Application of Basic Requirements, 24
3.6 Specification Requirements, 25
3.7 Compatibility, 26
3.8 Cost Comparison, 26
3.9 Commonality, 27
3.10 Cost Effectivity, 28
3.11 Delivery, 28
3.12 Summary, 29

4 Preparation of the Proposal 31

4.1 Technical Proposal Requirements, 31
4.2 Evaluation Factors, 34
4.3 Proposal Outline, 35
4.4 Proposal Language, 36
4.5 Introduction of the Proposal, 37
4.6 Technical Discussion, 38
4.7 Material and Hourly Effort Breakdown, 42
4.8 Scheduling, 43
4.9 Facilities and Personnel, 46
4.10 Description of Facilities and Experience, 47
4.11 Summary, 48

5 Contract Schedule 51

5.1 Description, 51
5.2 Analysis of Contract Schedule, 54
5.3 Summary, 59

6 Estimating of Costs 60

6.1 Introduction, 60
6.2 Cost-estimating Concepts, 61
6.3 Pitfalls of Estimating, 62

12.5   Make or Buy Decisions, 150
12.6   Assigning of Tasks, 152
12.7   Summary, 154

13   Configuration Management 155

13.1   Introduction, 155
13.2   Configuration Management Principles, 156
13.3   When Configuration Management Is Required, 158
13.4   Concept Formulation Phase, 159
13.5   Definition Phase, 159
13.6   Acquisition Phase, 160
13.7   Operational Phase, 161
13.8   Disciplines of Configuration Management, 162
13.9   Changes, 164
13.10 Summary, 166

14   Project Monitoring, Communication, and Decision Making 168

14.1   Communicating, 168
14.2   Program Status Communication, 170
14.3   Instructions and Direction, 171
14.4   Engineering Instructions, 172
14.5   Quality-control Instructions, 173
14.6   Drafting Instructions, 174
14.7   Purchasing Instructions, 175
14.8   Production Instructions, 176
14.9   Schedule and Cost Coordinator Communications, 177
14.10 Communication with the Customer, 178
14.11 Decision Making, 179
14.12 Summary, 180

15   Engineering the Equipment 182

15.1   Planning the Design, 182
15.2   Procurement of Data for Design, 183

15.3   System Design, 184
15.4   Block Diagrams, 185
15.5   Design of the Subsystem, 187
15.6   Design Check for Standardization, 187
15.7   Reliability, 189
15.8   Breadboards, 190
15.9   Packaging Design, 191
15.10 Drafting, 192
15.11 Testing, 194
15.12 Value Analysis, 194
15.13 Summary, 195

16   Digital Computer Applications 198

16.1   Computer Introduction, 198
16.2   Computer Characteristics, 200
16.3   Analysis of Computer Functions, 203
16.4   Summary, 206

17   Reliability and Maintainability 208

17.1   Background, 208
17.2   Reliability Concepts, 209
17.3   Implementation of Reliability in Design, 211
17.4   Selection of Components for Reliability, 213
17.5   Reliability Review, 217
17.6   Maintainability Principles, 218
17.7   Summary, 220

18   Production and Quality Control 222

18.1   Production of Prototype Equipment, 222
18.2   Production Planning and Control, 223
18.3   Quality Control, 227
18.4   Quality Control and Management, 228
18.5   Summary, 229

19 Test and Checkout 230

19.1   Test Objectives, 230
19.2   Test Criteria, 231
19.3   Types of Tests, 232
19.4   Scheduling of Test and Checkout, 237
19.5   Summary, 237

20 Supporting and Monitoring Items 239

20.1   Identity, 239
20.2   Engineering Reports, 240
20.3   Manuals, 242
20.4   Spare Parts, 243
20.5   Drawings, 244
20.6   Other Side Items, 244
20.7   Summary, 244

21 Follow-up 246

21.1   Postacceptance Considerations, 246
21.2   Finalizing the Contract, 246
21.3   Subsidiary Items, 247
21.4   Field Reports, 248
21.5   Unsolicited Proposals, 248
21.6   Summary, 249

Glossary, 250

Index, 258

Preface

Management of Engineering Projects is essentially a sequel to *Project Engineering*, which was published in 1965. Whereas the two books share a similar theme, the reader will find that most of the chapters in this edition incorporate major revisions that reflect updated management procedures and discuss disciplines related to the project engineer's task. In addition, new chapters have been incorporated that present such subjects as configuration management, legal contract points and the application of digital computers.

Readers of *Project Engineering* felt the title was restrictive since the subject matter embraced managerial disciplines that were beyond the strictly technical field the title implied. Because of the new and revised material and the added scope of technical and managerial responsibilities of the project engineer that are discussed in this text, the new title *Management of Engineering Projects* more accurately identifies the subject of the book.

In today's competitive technical arena, the decisions and influence of project engineers can determine whether a company will succeed in obtaining a contract award; when a contract is in effect, their actions can determine whether the engineering and profit objectives will be realized by the company. It is therefore essential that project engineers not only render sound engineering decisions but be familiar with and sensitive to the nontechnical issues they will confront. Much of the expanded and new material in the text discusses the issues project engineers must resolve.

The text of *Management of Engineering Projects* was designed to present chronologically the life cycle of a project from the time a company takes an interest in the project or procurement through its various phases, concluding with the acceptance and support of the end product of the program. The landmass simulator is used as the vehicle to illustrate the various principles and disciplines that involve the project engineer.

The broad scope of the project engineer's responsibilities has been recognized by many organizations. In an attempt to identify more accurately the role of the manager of a project, new titles for the project engineer were created, such as project manager, program manager, and project director. However, the head of a project, regardless of the title, is still charged with rendering engineering decisions in addition to executing managerial functions as discussed in the text. Therefore, *project engineer* is still accepted as the most accurate title for the head of a project.

It is hoped that this book will assist engineers, managers, and future managers to recognize the broad scope of disciplines that must be exercised by the manager of an engineering project.

Victor G. Hajek

Chapter One

Role of
the Project Engineer

1.1 Objectives and Disciplines

The project engineer is assigned by an organization the responsibility and authority to manage a technical project or program in a manner that will result in meeting the project objectives. Most projects involve a contract between two or more parties. However, an organization may embark on an "in-house" technical effort designed to achieve a particular end result, such as developing a new product or system that can be marketed, and a project engineer would be put in charge of such an endeavor. In this text, the type of project involving a contracted effort and the role that the project engineer fills in managing a contracted project will be discussed. To illustrate the many functions of project engineers and the wide scope of disciplines that they must apply, the case study of a government procurement of a Radar Landmass Simulator contracted from a fictitious company, Acmen Electronics Company, will be presented where applicable.

The Radar Landmass Simulator (RLS) is a training device comprised of electronic, optical, and mechanical subsystems. Engineering and design details will be presented only to the extent necessary to illustrate the technical decision-making responsibilities of the project engineer. Because the project engineer's responsibilities embrace nontechnical disciplines which are vitally important to the success of the program, much of the text will also cover nontechnical areas such as scheduling, contract law, etc.

The three basic objectives for which the contractor's project engineer is responsible are as follows:

1. Deliver a product that meets the requirements of the specification

2. Deliver a product that meets the requirements of the contract delivery schedule

3. Meet the company's profit objectives for the contract

The following are the prerequisites that must be satisfied if the above noted objectives can be expected to be achieved:

1. The company must have the particular engineering, production, and management resources required for meeting the project objectives. For example, a company would be hard pressed to design and produce equipment incorporating a digital computer if the company did not have experienced computer-programming personnel.

2. The company must have available adequate resources and facilities to meet the delivery schedule of the contract.

3. The terms of the contract, price, rates, etc., should be realistic and adequate to cover the company's cost to perform under the contract.

Project engineers have the dominant role in establishing the objectives to be achieved since they are responsible for the project from the time the request for proposals is analyzed through the equipment acceptance and often through the contract closeout effort. The achievement of the above objectives requires that they be proficient in a scope of disciplines covering managerial as well as engineering functions. The major disciplines and the functions for which the disciplines would be applied are summarized below.

1. *Engineering:* By its nature, a project-oriented task cuts across several areas of engineering effort. For example, a military weapon system might require expertise in such disciplines as electronic, electrical, and mechanical engineering. If the project was for a processing plant the required disciplines would include chemical, electrical, electronic, and mechanical engineering. In addition, specialists in engineering subgroups have evolved during recent years. For example, a digital computer engineer is generally an electronic engineer who has specialized in computer design or application. Project engineers cannot be expected to be expert in all the technical areas that a project requires, but they must be familiar enough with the basics of different technologies to consider alternatives that the engineering specialists may offer and be able to evaluate and make correct decisions in the various areas.

2. *Cost Management:* All the various areas for which project engineers are responsible can be ultimately reduced in most cases to one common denominator—money. If a particular engineering design approach proves to be in error, the necessary redesign would involve unanticipated expenditures of money. Or if the time to produce a product exceeds the planned schedule, an unanticipated expenditure of money would be necessary since the time that equipment remains in a plant, unshipped, consumes money.

The discipline of cost management, which is one of the prime responsibilities of project engineers, embraces the following functions:

Cost estimating
Cost accounting
Project cash flow
Direct and overhead rate controls
Incentive, penalty, and cost- or profit-sharing consideration
Cash-flow considerations

3. *Contract Law:* A project involving a buyer and a seller is almost always based on a legal agreement or contract between the participating parties. Project engineers, who are assigned the responsibility to have their employer's contractual obligation fulfilled, must be fully knowledgeable of the contract terms as well as sensitive to their implications. Normally both the procuring activity and the contractor have project engineers—both charged with the responsibility of fulfilling their employer's contractual obligations. As noted earlier, this text will deal primarily with the project engineer representing the contractor.

4. *Negotiations:* Most of the larger organizations employ specialists trained in negotiation psychology, tactics, procedures, and standards of conduct. However, the specialists rely upon the project engineers to assist in negotiation conferences and to provide detailed technical information. As a participating member of their company's negotiation team, project engineers must be aware of what or what not to say, when silence should be maintained, and when to speak up.

In smaller organizations, project engineers often are the company's key persons in negotiations. In either case, they represent a key factor in determining the success of their company's efforts in obtaining favorable contract awards.

5. *Scheduling:* A contract for the design, fabrication, and delivery of equipment for which the project engineer is responsible almost always involves complex scheduling problems. The proper phasing and scheduling of different types of effort required by the project are essential to the meeting of the contract delivery dates. Project engineers must therefore be familiar with the various engineering tasks, processes, and available resources which are necessary in order to execute the contract.

In addition, they must be able to apply the various tools and techniques that are available to implement effective scheduling of the project. Some of the management tools that might be used by the project engineer include:

PERT

Line of Balance (LOB)

Gantt Charts

1.2 Authority of Project Engineers

A company or organization will make the decision to pursue a particular program when the management officials decide that the benefits that will be realized by achieving the objectives warrant the risks and effort involved. Some of the benefits that would attract a company or organization are profit, the development of a new product that would result in profit, and the creation of a usable end item to serve a need.

When the decision to pursue a program as a project-oriented effort is made by management, the officials will select a project engineer who will be charged with the responsibility for achieving the program objectives and who will be provided with the authority to carry out these responsibilities.

The authority given to a project engineer is broad in scope and embraces the disciplines discussed in Section 1.1. Some of the more common areas of decision-making authority include the following:

1. Technical decisions
    a. Directing the design approach
    b. Selecting subsystems or components to be used
    c. Identifying the type and scope of tests
2. Commercial decisions
    a. Make or buy decisions
    b. Selecting or recommending subcontractors and vendors
3. Administrative decisions
    a. Selecting and assigning personnel
    b. Scheduling personnel, equipment, and other resources on the program
4. Monetary decisions
    a. Determining the expenditure of budgeted funds

1.3 Project Engineers' Responsibility to Management

As previously noted, the project engineers' responsibility to management is to achieve the objectives of the project or program. However, the one thing that management cannot tolerate is an unpleasant surprise, such as an unreported technical problem or an unforeseen expense. Therefore it is essential that project engineers report accurately, factually, and promptly all problems, errors, and potential problems.

In the case of a contractor's profit objectives, project engineers must maintain a close control and scrutiny of the funds that are involved on the program. Some of the monetary responsibilities that they have include the following:

1. Making judicious expenditures
2. Maintaining adequate cash flow into the program by meeting progress-payment milestones
3. Avoiding circumstances that might necessitate loans to keep the program moving
4. Keeping profit objectives visible
5. Avoiding errors

The carrying out of the project engineers' responsibility to management requires that they communicate the technical, delivery, and financial status of the program in a timely and frequent manner. The most critical requirement of any report to management is that it "tells it like it is." Any attempt to gloss over actual or potential difficulties is bound to be exposed at a later date, almost always too late for effective remedial action.

1.4   Summary

Project engineers are given the responsibility for successfully meeting the technical, delivery, and financial objectives of the program assigned to them. In order to carry out their responsibilities, they are given the authority to prosecute their functions.

In most cases, the program that the project engineer monitors includes a contract between the procuring organization and the contractor. Project engineers would be assigned the responsibility for carrying out the customer's objectives and/or the contractor's objectives. Regardless of whether project engineers represent the customer or the contractor, they must have a grasp of the fundamentals that cut across a broad spectrum of disciplines, which would include the various branches of engineering, cost management, contract law, procurement regulations, negotiation techniques, scheduling procedures, management presentation, and other areas of management. In the final analysis, the management officials of the organization which the project engineer represents depend upon the project engineer for implementing sound decisions, making accurate and factual reports to them, and successfully carrying out the program objectives.

Chapter Two

The Procurement Project Initiated

2.1 Request for Proposal

The Acmen Electronics Company first becomes officially involved as a potential offeror in a procurement when the company receives the Request for Proposal (RFP) for the RLS.

The RFP contains the information necessary for the offeror to prepare the technical proposal and the price quotation. The primary documents comprising an RFP include the following:

1. *Covering Letter:* Forwards the documents relating to the procurement and gives a general description of the items to be purchased, the type of contract solicited, and any other basic information of interest.

2. *Specification:* Stipulates the performance and physical requirements for the articles under procurement.

3. *Contract Schedule:* Indicates line items to be provided, delivery and other milestone dates, contract clauses, incentives and/or penalties, and any other contractual terms that are applicable to the proposed procurement.

4. *Technical Proposal Requirements (TPR):* Identifies the specific information that must be provided by offerors in their proposals. The details of the TPR are derived from the specification and serve to guide

offerors as to the content of their proposals, thereby avoiding proposal material which the user neither desires nor intends to evaluate.

5. *Cost Breakout (for negotiated procurements):* Identifies the depth of cost detail that is required in the cost portion of the proposal. The detail can range from a single number for each line item that is listed in the schedule to a breakdown of cost for each kind of discipline (e.g., mechanical engineering, testing, and sheet metal processing). The cost breakout would normally include labor rates, overhead, G&A (General and Administrative) rates, and profit, among other things.

6. *General and Administrative Clauses:* Information relating to the procuring organization's policies. The offeror must indicate specific concurrence with the clauses in the proposal submitted. Typical of such information might be patent rights, prohibition of prison labor, and employment of minority personnel.

7. *Data:* Supplementary information to aid the offeror in identifying the articles to be procured. Data could be in many forms, including drawings, sketches, and publications.

2.2 Participation Decision

The preparation of technical and cost proposals, attendance at preproposal meetings, involvement in negotiations, etc., require the commitment of significant resources and money by a company. Since only one company will be awarded the contract, the funds and effort represent a total loss to companies that have not been successful. Therefore, the decision to participate or not participate in a certain procurement must be based on careful analysis of the circumstances surrounding it and an evaluation of the risks that the company must be willing to accept in the interests of being successful in receiving a contract award.

The individual to function as the project engineer on a program is not officially designated until the decision to participate is made by management. However, the project engineer–designate will participate in preproposal discussions and provide inputs as appropriate. The points that must be considered in deciding whether or not to participate in a procurement were alluded to in Section 1.1 and are summarized as follows:

1. *Technical Resources:* Does the company possess the type and depth of engineering technology that the design of the items under procurement requires?

2. *Facilities:* Are the space, equipment, testing and other facilities adequate for the program?

3. *Work Load:* In event of a contract award, would the company be able to handle the work-load demands in addition to the existing work load?

4. *Competition:* Does some other company have a significant advantage due to prior experience, expertise, patents, proprietary information, or political influence?

5. *Delivery:* Are the delivery requirements reasonable and within the company's ability to perform?

6. *Risks:* Does the procurement pose any significant risks to the company in such areas as technical difficulty, penalties on performance or delivery, or type of contract?

If, after consideration of the points identified above, the company decides to budget funds, prepare a proposal, and participate in the preaward effort, the project engineer will be assigned to the program and the proposal team will be organized. From that point on, the project engineer will be responsible for making the proper decisions through all the phases of the program. For the case of Acmen Electronics Company, it is stated that the management officials made a thorough analysis and determined that the company met all the above noted criteria. As a result, the decision was made to have the company commit the funds for proposal preparation and compete for the contract award.

2.3 Technical Considerations

The technical role of project engineers begins in earnest when they start to analyze the procurement specification and the TPR documents. They must be intimately cognizant of the specification details to understand what must be designed, developed, and fabricated; and they must know the details cited in the proposal requirements document in order to effectively address, in the technical proposal, those points which will be evaluated by the procuring officials. It is at this point that they must rid themselves of any preconceived ideas as to what is required and carefully study the specifications and proposal requirements with objective and open minds. All too often large expenditures of effort and money are made in preparing a proposal which reflects what the project engineer would like or is best qualified to offer rather than what is required in the specifications. Unless the proposal written or directed by the project engineer addresses the points cited in the TPR document and provides information that will be judged superior to what the competition offers, the contract could be lost and the project engineer will have no project to direct.

Specifications vary from design documents which define the required object in detail to performance documents which merely describe the end use of the product without defining the method, means, or design concepts to be used. A design specification provides engineering design information so that the ultimate contractor would not be required to create an original design. The contractor would have to possess the ability to interpret and translate the information of the drawings and design documents into the hardware. A performance specification, however, requires creative engineering effort. The more general the performance specification, the greater the imagination that must be applied for the successful execution of the program. Since the role of the project engineer is critical and vital in administering contracts based on performance specifications, this text will confine itself to analyzing and resolving the common problems that would be encountered in procurements based on such specifications.

2.4 Performance Specifications

During the course of a year, many different kinds of performance specifications are written for equipment of many types, and thousands of companies submit proposals with hopes of obtaining contract awards. Except for the specific technical knowledge required, the role and functions of the project engineer are, broadly speaking, the same for any project. One major field of procurement in which performance specifications are widely used is the electromechanical and electronics field. Although the case discussed in this text will deal with electronics, the principles discussed and the logic on which the decisions are made apply to practically any field of endeavor in which a project engineer may function.

In the following text the various principles that are illustrated, the analysis of problems, and the development of the reasoning behind decisions will be discussed from the point of view of a project engineer of the contractor, Acmen Electronics Company, and the company's functions in seeking a contract award and subsequently delivering to the United States government the RLS. The customer will be the government, since the greatest number of project engineers are involved with government procurement. The principles, however, would be applicable to procurements by commercial customers and corporations.

Upon completion of the review of the specification, the project engineer should prepare a summary of the major and significant details of the project requirements. A specification summary of the RLS is presented below. Preparation of a summary will serve to identify and keep visible the major requirements of the project for both the project engineer and management officials.

SUMMARY MAJOR REQUIREMENTS
SPECIFICATION NO. 1001   DATED 1/15/77
Radar Landmass Simulator—DEVICE 5A1
   1. Scope
      a. Design, develop, construct Radar Landmass Simulator—Device 5A1.

   b. Simulated radar return from any preselected geographical area is required.

   c. Position and movement of aircraft to be simulated.

   d. No specific design approach is specified. Each of the offerors to propose their best approach.

   e. Requirements for performance accuracies, reliability, size, and testing are specified.

   f. Sixty-five week delivery required.

2. Detailed Requirements

   a. AN/APQ-28 airborne radar to be simulated.

   b. Simulated parameters

   (1) *Altitude*   0–20,000 feet.

   (2) *Speed*   250–600 knots.

   (3) Pulse repetition rate, antenna beam pattern, pulse width, and other details identified in Specification MIL-E-784K for AN/APQ-28 radar.

   (4) Device to simulate shadows, aspect angles, reflectivity characteristics.

   c. Resolution of simulated radar—not greater than 250 feet.

   d. Size of simulated area—1,500 miles by 800 miles.

   e. Accuracies of simulated radar

   (1) *Bearing*   ±2 percent.

   (2) *Range*   ±5 percent

   f. Size of simulation system—7 feet high by 16 feet wide by 4 feet deep.

   g. Weight—no restriction.

   h. Flexibility—design should be flexible enough to permit modified radar characteristics or additional terrain features to be simulated.

   i. Testing of simulator

   (1) *Environmental* In accordance with specification MIL-T-17113.

   (2) *Performance* As necessary to certify compliance with Simulator Specification N1001. Detailed tests to be established by procuring activity within 6 months from contract award date.

   j. Simulator reliability equipment

   (1) *Specified MTBF (Mean Time Between Failures)*   100 hours.

   k. Government Furnished Equipment and data to be provided.

3. Referenced General Specifications

MIL-T-17113	Tests, Shock, Vibration and Inclination (for electronic equipment); General Specification
MIL-E-16400	Electronic Equipment Naval Ship and Shore; General Specification
MIL-E-784K	Radar Set AN/APQ-28; General Specification
NTEC-3111-962	Standards for Engineering Design Reports
NTEC-423-912	Standards for Engineering Drawings
NTEC-512-352	Requirements for Maintenance Manuals
NTEC-3102-100	Military Training Devices; General Specification

2.5   Analysis of Specification Requirements

The completion of the summary of the performance specification high-lights what is required by the procuring activity but leaves the design approach open to the discretion of the offeror. The RLS is described by an essentially pure performance specification. Very often a hybrid performance/design specification is used which may reduce the design-approach options available. For instance, the specification cited herein could have required an approach based on the use of digital computers, thereby restricting the wide range of design approaches that offerors might wish to consider.

Throughout the program, the project engineer must keep in continuous focus the major requirements to avoid having irrelevant factors influence a technical or administrative decision.

In any contemplated task, the interested party has to determine how best to meet the task requirements with the resources available. The same principle has to be considered by project engineers as they review the requirements of the new procurement.

After carefully studying the specification and all referenced documents and having satisfied themselves that they fully understand the specification requirements, the project engineers and their staffs must select the technical approach which best satisfies the following approach criteria.

1. It is the approach that offers the best design for the equipment under procurement and is consistent with the specification requirements.

2. It is the approach that is best suited to the experience, facilities, and capabilities of the company.

3. It is the approach that will permit the company to offer the lowest cost and best delivery for the procurement.

In some cases, no one approach would satisfy all the above requirements. When such a situation occurs, it is mandatory that a carefully thought-out compromise be selected that offers the best chance of obtaining the contract, meeting the contract requirements, and making a profit.

2.6   Technical Approach

The RLS that is described in the specifications that accompanied the RFP suggests several possible design approaches. The objective analysis of each different design approach will enable the project engineer to converge on a decision regarding the one design that best satisfies the three approach criteria discussed in Section 2.5.

The various technical designs that are possible and must be considered can be identified as follows:

1. Use of three-dimensional terrain models with sound sources
2. Use of three-dimensional terrain models with light sources
3. Use of transparencies (incorporating geographic area) having various shades of gray with light sources
4. Use of multicolor transparencies with light sources
5. Use of digital computers

The above basic techniques must be analyzed to determine how well each meets the major contract requirements which are derived from the specification.

In order to illustrate the analytical processes that the project engineer must perform in deciding on a technical approach for the procurement in question, a basic description of each design is given with an analysis of how well each approach meets the specifications, the company capabilities, and the cost objectives. In actual practice, the project engineer will already be familiar with each design or will be required to obtain the information from sources available for this knowledge.

The first possible approach is based on a three-dimensional model using sound sources. This system simulates radar landmass returns by transmitting directional audio sound pulses at the terrain model and sensing the echo returns. The resultant signals are processed and displayed on the simulated radar scope. The terrain model must be immersed in water to facilitate the simulation of speeds of propagation of radar pulses. The sound source simulating the radar antenna is driven over the terrain model in the three (X, Y, Z) axes by means of a gantry. The model scale, rates of movement, and other factors must be precisely designed to simulate geographic distances, rates of radar propagation, and other effects.

A deduction that would be immediately obvious to the project engineer is that this approach could not meet the requirement of size since the smallest possible terrain model scale of 250,000 : 1 would require a model of about 36 by 20 feet, exclusive of the electronic consoles and other systems that would be necessary. The three-dimensional terrain system presents several problems for meeting the environmental tests because of unwieldy size. Also the flexibility requirement for problem area changes would be very difficult if not impossible to meet. The inherent design features which would not permit compliance with the specification would eliminate any further consideration of this particular design. The ability of the company to deliver at an economical price a three-dimensional model using sound sources is in no way a consideration because of the specification noncompliance problem.

The second approach involves a three-dimensional model using light sources. The basic principle of this system is identical to the system using sound sources except that light is the energy source and water is not used. The size, weight, inflexibility, and other limitations are the same for all three-dimensional model systems. Because of the inherent design of such systems, the project engineer can eliminate the three-dimensional models from any consideration since such systems obviously would not be able to meet the specification requirements.

The third method is referred to as the Gray Scale Transparency System. Systems utilizing transparencies of gray shades incorporate a moving pinpoint light source and a light-sensitive pickup tube. Different shades of gray on the transparency represent different land elevations and terrain objects such as bridges and buildings. The light impinging on specific areas of the transparencies simulates the radar energy from an aircraft illuminating the terrain areas as the aircraft traverses a mission course. The various shades of gray on the transparencies modulate the amount of light going through the transparency that is detected by the pickup tube. The light so detected represents specific terrain intelligence that is processed and displayed on the simulated radar scope.

The transparencies showing a geographic area can be scaled to a 5,000,000 : 1 size so that the transparency size of 1.8 by 1.0 feet and associated optical and electronic gear present no difficulty in meeting the specification size requirement. In like manner, the compactness and small size of this system offer inherent capability of permitting a design that will meet the specification requirements for environmental tests, flexibility, area of operation, and other requirements. Since the project engineer's company, Acmen Electronics, has the experience and resources to produce a RLS incorporating the transparency design, the approach would make the procurement very promising for the company.

The fourth approach, referred to as the multicolor transparency technique, is similar to the gray scale method. The terrain information is coded in varying intensities of three colors on the film transparencies. The pinpoint light source is modified by the colored transparency. The resulting light that has penetrated the transparency is filtered and a detector for each of the three colors senses the resulting colored light intensity which represents part of or all the terrain information of a particular point. This information is processed for eventual display on the simulated radar scope.

The physical properties of the multicolor system are very similar to the gray scale system so that this system offers promise of providing an approach to propose for the procurement.

The digital computer approach is the fifth possible method that must be considered. Instead of storing the minute and detailed terrain data on some physical map medium such as a three-dimensional model or a transparency, the terrain data are programmed for the digital computer. The geographic area of 1,500 by 800 miles must be divided into discrete points, each of which has its individual characteristic that affects the radar returns such as elevation, type of topography, type and configuration of objects or structures, terrain features, and other information that is coded and stored in the memory of the digital computer. When the aircraft is flown over the area, the simulated radar beam sweeps over a particular sector of the gaming area and the returns are displayed on the trainer radar display. The synthetic returns on the display must realistically embody the major characteristics of the returns of the area that is scanned.

In the digital system the terrain characteristics of a large number of points must be stored. If the area specified is to be divided into a series of grid lines 100 yards apart, each intersection representing a terrain point to be stored, the memory unit must have a capacity of approximately 4.8×10^8 words. The 2,000-square-mile area that the AN/APQ-28 radar antenna beam can illuminate at a typical altitude would be comprised of 8.0×10^5 grid points.

The next fact to be established is the rate at which stored data must be retrieved. An aircraft flying at 600 knots travels at about 300 yards per second and therefore traverses each line of grid points in about one-third second. The speed at which the aircraft travels over the simulated terrain is one parameter that must be considered to calculate the rate at which the data must be processed.

In addition to the speed of aircraft travel, the radar pulse speed, its range, the antenna sweep, and other functions must be handled by the digital computer. Whereas the requirements for computer computation speed, storage capacity, and other requirements of the digital computer are within the state of the art, nevertheless special programming techniques, methods of selective sampling, and special circuitry indicate that a major research effort would be required which would involve many risk factors and unpredictable development time.

From the above analysis of the major requirements and the five possible technical approaches that were identified, the project engineer would conclude that the following three approaches are technically feasible for meeting the specification requirements of the RLS:

Gray Scale Transparency System
Multicolored Transparency System
Digital computer design

After the primary determination as to which design approaches are technically feasible, the project engineer must make further screening to find which approach best meets the criteria described in Section 2.5

above. The second screening for the selection of the design approach that will be proposed by Acmen Electronics would be established as the result of the following analysis.

1. The approach that offers the best design for the equipment under procurement and is consistent with the specification requirements: A survey will show that some work has been done in all three areas but that the Gray Scale Transparency System might be considered to be furthest advanced. This does not necessarily mean that the Gray Scale Transparency System is best, but at least its adaptation offers a contractor the best chance for success. It also means that the adoption of this system would require the least amount of research and development work.

The project engineer's survey would reveal that even though the Gray Scale Transparency System may have some inherent limitations, further study of the specification requirements and the gray scale transparency design would indicate that all the specification requirements could be met.

2. The approach that is best suited to the experience, facilities, and capabilities of the company: The transparency approaches (either gray scale or multicolored) require personnel highly skilled in optics engineering, photographic technology, and electronic circuitry. The digital approach requires digital computer engineers, programmers, and electronic engineers. The Acmen Electronics Corporation can be classified as having the following qualifications and experience: (1) electronic circuitry design, (2) optical design, and (3) photography (black and white only). Of the possible areas discussed, those in which the company lacks experience are color photography and digital computation and programming.

3. The approach that will permit the company to offer the lowest cost and best delivery for the procurement: In view of the experience, personnel, and facilities of Acmen Electronics and in view of the fact that the Gray Scale Transparency System offers a proved technical approach, this particular design approach holds promise for the offering of the lowest price and the best delivery.

To conclude, the above analysis of the project engineer would lead to the choice of the Gray Scale Transparency System as the technical approach for the procurement in question. The choice is based on the following:

1. Whereas the Gray Scale Transparency System has been developed to a point where an experimental model is in existence, other systems are in different states of perfection.

2. The company has no special experience in the field of color photography or related fields, so that it would not be in a position to offer the multicolored technique and thereby enhance its competitive position.

3. The use of the gray scale transparency technique for the simulator is not only best suited for the experience and facilities of the company, but also offers the best avenue for the company to permit a bid of the lowest price of the three systems considered. In like manner, the shortest delivery is also facilitated by offering the gray scale transparency approach.

Thus, even though the multicolored or even the digital system offers promise for a greatly superior landmass simulation system, practicality dictates that the project engineer choose and offer the gray scale approach in the technical proposal.

2.7 Summary

The RFP officially establishes the identity, schedule, and desired terms for a proposed contract. The RFP is comprised of various documents including a covering letter, specification, proposed contract schedule, TPR, cost breakout forms, special clauses, data, etc.

The preparation of a technical proposal and participation in a contract solicitation requires the commitment of funds and resources by each offeror. It is therefore essential that offerors evaluate the RFP and their capabilities to determine if their participation in the solicitation justifies the effort that would be required. Some of the factors that must be considered in arriving at the decision include an analysis of the company's technical resources, facilities, work load, competition, delivery schedule, and inherent risks in the execution of the desired contract.

If the decision to participate in the solicitation is made, the project engineers must analyze the specification requirements to determine the design approach that (1) best satisfies the specification, (2) is compatible with the capabilities and facilities of their company, and (3) offers the best approach for meeting the profit and delivery objectives of the company.

The analysis of the RLS was made to demonstrate how various technical approaches would be considered in deriving the design approach that would be ultimately adopted for the proposal and, if a contract award were received, the design of the equipment that the specification describes.

The analysis of the possible design approaches was described as involving the following two phases:

1. *Primary analysis:* to determine the design approaches that would satisfy the specification requirements

2. *Secondary screening:* to identify the design approach that would be most compatible with the resources, experience, and facilities of Acmen Electronics Company

PROBLEMS

The ABC Electronics Corporation, specialists in optics and cameras, has received a RFP for equipment which is to plot the movements of ten separate targets on a two-dimensional display, 36 by 36 inches. The essence of the specification is as follows:

> The Contractor shall design and fabricate a two-dimensional display system that shall reproduce the course and speeds of ten separately moving targets. The movements of each target shall be shown as a single line trace, and each trace shall be a different color. The accuracies of each trace shall be within 2 percent in course and speed. Each target shall be activated from a 60-cycle, 110-volt servo system. The activating servo system signals will be supplied from the main computing system, which is presently installed and operating at the government installation site.

The project engineer's analysis of the specification and contract requirements reveals that either of the following approaches would meet the specification requirements:

1. A pen-and-ink recording system in which each pen is driven by servos to inscribe the target trace.

2. A projection system in which some servo-driven instrument inscribes a line or trace on an opaque sheet of glass. The projector light source will project the target trace on a screen for viewing.

Discuss the logic for choosing the approach which would offer the ABC Electronics Corporation the better chance for contract award and for successfully completing the contract.

Chapter Three

Proposed System Design Considerations

3.1 Equipment Systems and Elements

The design approach for the procurement represents the key decision that will determine whether or not the company will be successful in winning a contract award.

During the solicitation or contract definition phase, the primary concern of the project engineer is to draft a technical proposal that is responsive to every point cited in the TPR document. Since the details of TPR were derived directly from the specification, the establishment of the design-approach details would be used in the equipment design itself, if the company is awarded the contract.

The systems engineer is primarily concerned with the overall engineering design that would best satisfy the requirements of the specification at the least cost. Because the parameters of cost and time are prime considerations, the systems engineer strives to avoid identifying subsystems that involve extensive development. In effect, for any project, the systems engineer attempts to establish a design comprised of subsystems that may have been previously engineered or subsystems that can readily be engineered without becoming involved in major development effort or state-of-art considerations.

The systems engineers, who work closely with the project engineers, become involved with a project during its earliest phases. The results of their effort become the basis for the Work Breakdown Structure, which is a means by which the equipment subsystems are identified as tiers of design detail. A Work Breakdown Structure is usually derived in the following sequence:

1. Translating performance objectives into a system technical approach

2. Dividing the system into tiers of design similar to what is depicted in Figure 3.1

3. Selecting feasible subsystem designs that are compatible with each other and that will satisfy specification requirements

4. Reviewing and revising the system design approach so as to optimize from the point of view of cost, required performance, and schedule

The project engineers assign the task of describing each of the subsystems identified in the Work Breakdown Structure to cognizant engineers of the proposal team. Their subsequent task would be to collate the material that is submitted, edit the information to include only what is required by the TPR document, and organize the material into a proposal that hopefully will serve to convince the project engineer of the procuring agency that the Acmen Electronics Company's technical approach will best satisfy the procurement specification.

3.2   Tiers of Design Detail

Any system can be broken up into tiers of design detail (or a Work Breakdown Structure). In the case of the landmass simulator design, the tiers of design detail are shown in Figure 3.1. The first tier of detail identifies the major systems which make up the equipment. The higher levels of management to whom the project engineer reports would be primarily interested in the detail reflected by tier I of the design.

Tier II is a further breakdown to the RLS and, in general, depicts the major subsystems derived from tier I.

Tier III relates to details assigned to different design engineers of groups. In effect, tier III represents the detailed Work Breakdown Structure necessary for the effort that the design of the RLS involves.

Project engineers would be primarily concerned with the first and second tiers of technical effort as far as their functions are concerned. If any particular problem developed, they might actively participate in the third or lowest tier of effort, but only to become knowledgeable enough with the details associated with these areas in order to make some basic decision or to make a complete report to higher management.

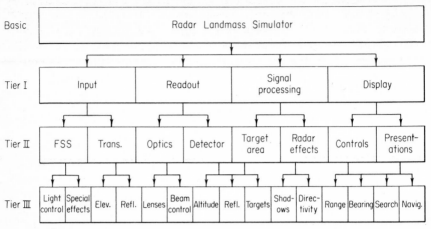

FIG. 3.1   Radar Landmass Simulator, tiers of design detail.

3.3   Preliminary Block Diagrams

The sequence of functions of the various subsystems of a piece of equipment can be illustrated by means of a block diagram. Thus, the use of a block diagram serves as a very effective tool for understanding any complex system. It serves to portray graphically the flow of signals and the interrelation among subsystems and serves as the basis for further detailing of the overall design. Effective technical proposals will invariably make liberal use of block diagrams.

The block diagram for the RLS which reflects the tier I detail shown in Figure 3.1 is illustrated in Figure 3.2. The degree of detail reflects only the major subsystems and portrays the design in its broad terms.

The block diagram representing tier II detail of the RLS is shown in Figure 3.3. The tier II detail, in general, represents what would be presented in a technical proposal. A few of the elements of the tier II block diagram will be analyzed and discussed to illustrate the logic and steps that the project engineer must use in arriving at decisions as to the simulator subsystem design that is to be proposed. No attempt will be

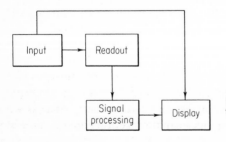

FIG. 3.2   Radar Landmass Simulator block diagram (tier I).

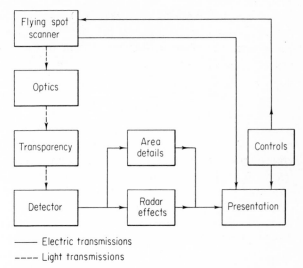

FIG. 3.3   Radar Landmass Simulator block diagram (tier II).

made to describe how the project engineer arrives at technical decisions relating to all the elements or blocks of Figure 3.3 since we are only concerned with the logic of the deductions and not with the design of the landmass simulator itself. The method of deduction could be applied to any project or type of equipment. The elements of Figure 3.3 which will be analyzed are the Flying Spot Scanner (FSS) and the optics system.

3.4   Factors Affecting System Approach

In considering the elements of a system, project engineers must select the design that best satisfies those basic requirements previously discussed, namely the following:

1. Specification
2. Company resources and capability
3. Cost
4. Delivery

In addition, during the course of their design management functions, project engineers must assure that the necessary compatibility among the various subsystems that make up the trainer design is achieved.

The following elaborates further on the basic requirements under discussion.

1. *Specification Requirements:* The performance specification describes the major item of procurement in general terms only. The description relates primarily to the end product, describing what is required and providing the detailed performance requirements such as

accuracies, reliability figures, and other similar information. In addition to the description of the end product, the specification will cite other documents which set forth requirements for components, environmental tests, and similar standards that are generally applicable to all types of equipment that might be procured by a company or a government agency.

The project engineer, in selecting a design approach, is usually faced with a paradox: offering the optimum design for the least amount of money. Careful consideration must be exercised when faced with a choice of a superior design at a higher cost versus a marginal design at a lower cost. All the factors, including an accurate evaluation of how the customer's project engineer will react to a proposed design, must be considered in making a choice. From the contractual point of view, the primary objective of the technical proposal is to support the premise that the equipment that the offeror describes will meet the requirements of the specification. Here "requirements" means the legal connotation, "minimum requirements." Offerors may find that they are penalizing themselves by offering equipment which may exceed the requirements of the specification but finding that the equipment is more costly to build. Project engineers must therefore be realistic and propose the least expensive system that satisfies the specification and other contract requirements. If they are convinced that the customer would be interested in the advantages that a more expensive design offers, an alternate proposal can be submitted. Such an alternate proposal must prove to the user that the advantages to be realized from a superior design warrant the higher cost. Regardless of the project engineer's decision, the fundamental fact that must be constantly recognized is that any design that is offered must meet the requirements of the specification.

2. *Company Resources and Capability:* The broad scope of various technical fields and the stringent requirements that have been imposed on the equipment that is engineered and fabricated in the vast number of individual fields have forced different companies to specialize in different fields. In the electronics field for instance, companies specialize in such areas as power supplies, antenna systems, receivers, and analog to digital converters. A company which did not have the resources of experienced personnel, equipment, and facilities would be at a great disadvantage in competing in an area which is new to them. Project engineers must therefore choose the technical approach which will satisfy the specification requirements and which will be in a technical area that their company can comfortably handle.

3. *Competitive Cost:* The proposal bid price must consider the following points: (a) the project engineer's cost breakdown comprising engineering, manufacturing and material costs, overhead rates, and profit

objectives; (b) contingencies for risk areas and other unknowns; and (c) management revisions based on the competitive environment, existing work load, potential for additional similar contracts, and other company policy considerations.

The success of a procurement award depends to a large degree upon realistic and accurate estimates of the cost breakdown submitted by the project engineer. An overcautious estimate will lead to a high bid, which could result in a lost contract. An overoptimistic cost breakdown will lead to a low figure, which could result in a contract award but end up as a money-losing effort for the company.

4. *Delivery:* All procurements are tied down to a schedule and to delivery dates of specific items. The type and urgency of a procurement will determine the contractual penalties and/or incentives related to meeting the delivery dates. As far as the design and engineering effort are concerned, project engineers must propose a design approach that is not only most cost effective for their company, but that also presents no problems regarding the meeting of the contract delivery schedule. In deciding upon the various approaches that might be chosen, project engineers must evaluate the areas that could involve special engineering or development and provide for the probability that delays will be experienced in perfecting those areas of engineering risk. The company's internal schedule of engineering effort, manufacturing time, and testing must also reflect the necessary scheduling contingencies.

The delivery schedule of any procurement is of prime importance to the seller as well as to the customer. Customers must recognize that the development of any new design or approach is time-consuming. They must weigh the expected advantages of the new approach, the chances for its success, the cost, and the time it is expected to consume in development. If the situation requires urgent delivery, then the customer must settle for equipment based on existing or proved design concepts. On the other hand, a requirement may exist which necessitates a design breakthrough based on some creative engineering and development effort. In such a case, the delivery must assume a secondary role to the development effort.

In some cases, usually associated with military procurements, a design breakthrough is necessary on an urgent basis. To achieve both objectives requires unusual effort and expenditures of money. In other cases, several different design approaches to achieve an objective are conducted simultaneously by different organizations or contractors with the hope that one approach might prove successful within the schedule requirements. This type of procurement is relatively rare.

Another reason why the delivery schedule is important to the contractor is that contracts often incorporate penalty and incentive clauses that are tied down to delivery. Also, as long as the equipment being provided remains in the contractor's plant, it accumulates costs to the contractor—even if no work is being done on it. Therefore, it is to the economic advantage of the contractor to deliver as soon as possible.

3.5   Application of Basic Requirements

The various factors that have been described will be discussed as applicable to two of the system elements, namely, the Flying Spot Scanner (FSS) and the optics system shown in Figures 3.3 and 3.4. The discussion of different systems or any other technical design detail is presented in this text solely to demonstrate principles or procedures that must be pursued by the project engineer and not to serve as an engineering text in those technical areas.

One of the important requirements that the project engineer had to recognize in the specification summary presented in Section 2.4 is that the trainer must be capable of achieving a resolution of at least 250 feet. In other words, the display of the simulated radar system must be able to distinguish two separated objects located as close as 250 feet from each other. If the two objects appear merged on the display, then the resolution requirement has not been satisfied.

Since the design approach utilizing a FSS was chosen, the project engineer must have sufficient technical knowledge to appreciate that spot

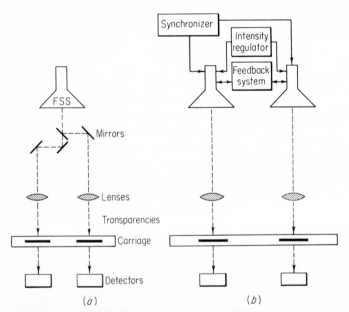

FIG. 3.4   Alternative approaches for FSS and optical systems.

size of the FSS must be sufficiently small so that the spot would not cover a diameter area greater than 250 feet on the scaled transparency. The project engineer must therefore calculate the map scale that can be used on the transparency and the largest spot size that can be tolerated in the FSS unit. Based on a number of important considerations such as the limitations of transparency gantry movements, transparency scale size optical system, and performance characteristics of commercially available scanners, the project engineer would conclude the following:

1. A transparency scale of 5,000,000 : 1 is to be used.

2. Flying Spot Scanners having a spot size of 0.0005 inches diameter (½ millimeter) are commercially available.

3. The scanner projecting a ½-millimeter spot size used with the 5,000,000 : 1 transparency will provide a resolution of 208 feet, which is well within the specification requirement of the 250-feet resolution limitation.

The above describes the type of analysis the project engineer would have made in choosing one of the commercially available components to be used in the simulator system. The choices of components are extremely important since the functions of the components and subsystems determine whether the equipment itself will perform in accordance with the specification requirements.

Having selected the FSS to use, the project engineer must determine whether two synchronized units or one unit with some means of optical system to split the FSS beam is best. The alternative approaches are shown in Figure 3.4. The considerations leading to a choice of scanner and optic subsystems are described below.

3.6 Specification Requirements

The specification requires accuracies of bearing and range for the radar presentation which necessitate that the beams of light scanning both the elevation and the reflectivity transparencies be synchronized to a high degree to guarantee that each element of information on each transparency be sensed simultaneously. Also, the ratio of light impinging on each transparency must be precisely maintained at all times.

With two independent scanners, an elaborate system of highly accurate synchro systems and feedback circuits must be incorporated in the system to achieve the synchronization and light-control requirements noted above, as shown in Figure 3.4. An optical system that splits the beam of a single light source can readily be utilized incorporating an optical arrangement which can automatically maintain the required synchronization and relative levels of light intensity for both transparencies.

The project engineer's comparison of the alternate design approaches shown in Figure 3.4 would indicate that the two-FSS system involves, in addition to the two light sources and their activating and control circuits, an elaborate system of servos and feedback circuits which must function in a highly precise and reliable manner. One basic criterion of good design is simplicity both in the number of elements and in the complexity of circuit. The two-scanner system approach requires three electronic circuits in addition to two optics systems.

The single-FSS system requires a somewhat more complicated optics system because of the beam-splitting arrangement but requires no special electronic elements and systems. From the basic design point of view the split-beam scanner is superior to the two-scanner system.

3.7   Compatibility

In order that the various subsystems of a piece of equipment work effectively as an integrated unit, all subsystems must be compatible with each other. For example, a system performing its functions exclusively by digital techniques would have greater subsystems compatibility than a hybrid system consisting of analog as well as digital systems. The analog subsystem cannot directly interface with the digital subsystem without converting analog outputs to digital or digital to analog, as the case may be. Wherever technically and economically feasible, compatibility among subsystems is a necessary goal.

In comparing the two design approaches for the FSS and optical systems in Figure 3.4, the project engineer would recognize that the necessity for having the synchronizer, intensity regulator, and feedback subsystems adds an extra dimension of complexity to the design shown in Figure 3.4*b* and introduces some elaborate circuits that would not be required by the system in Figure 3.4*a*. The subsystems in 3.4*a* are more compatible with each other and the straightforward design approach greatly simplifies the engineering, fabrication, testing, and operation of the overall system.

3.8   Cost Comparison

In considering the various design options that are available, the project engineer must also keep in mind the relative costs of the different approaches. A rough order of magnitude of costs for the two systems shown in Figure 3.4 could be summarized in Table 3.1. Based on the relative estimated costs for each design approach, which indicates the simple scanner design to be about 60 percent of the cost of the dual-scanner system, the project engineer has another basis for favoring the design shown in Figure 3.4*a*.

TABLE 3.1 Comparative Costs for Flying Spot Scanner and Optical System Designs

Single Scanner Approach (3–4(a))			Dual Scanner Approach (3–4(b))		
Material:					
One scanner tube		$ 500	Two scanner tubes		$ 1,000
Split-beam optics		2,500	Two optical systems		2,000
Transparency drive		2,500	Transparency drive		2,500
Detector system		2,000	Detector system		2,000
			Beam synchronizer		3,000
			Intensity regulator		2,500
			Feedback system		2,000
Total Material		$7,500	Total Material		$15,000
Engineering:					
FSS system	150 hrs	$ 3,000	FSS system	200 hrs	$ 4,000
Optics	300 hrs	6,000	Optics	100 hrs	2,000
Carriage	200 hrs	4,000	Carriage	200 hrs	4,000
Detectors	150 hrs	3,000	Detectors	150 hrs	3,000
			Synchronize	150 hrs	3,000
			Feedback	150 hrs	3,000
			Intensity	150 hrs	3,000
Total Engineering		$16,000	Total Engineering		$22,000
Manufacturing:					
Fabrication	400 hrs	$ 6,000	Fabrication	700 hrs	$10,500
Assembly	300 hrs	4,500	Assembly	600 hrs	9,000
Total Manufacturing		$10,500	Total Manufacturing		$19,500
Total Estimate		$34,000	Total Estimate		$56,500

3.9 Commonality

The word commonality was originated by the Department of Defense in discussing the justification for award of a major procurement that took place in 1963. It crystallizes a factor inherent in the utilization of a multitude of similar equipments by a user (such as the United States Navy using squadrons of identical aircraft), but because of its nebulous character, it was difficult to identify and therefore rarely considered in evaluations of proposals.

Commonality is a factor which reflects the percentage of the components, modules, and other elements of a piece of equipment which are interchangeable with existing equipment of a different type. For instance, the Army may have in operational use a search radar using a particular cathode-ray tube (CRT) for the display of intelligence. For a procurement of an entirely different type of radar, company A may propose using a CRT identical to the search radar in operation, but company B may propose a completely new tube. The table that follows summarizes the various commonality factors and indicates why company A would be preferred to company B, all other factors being equal.

Commonality factors	Company A	Company B
1. Savings due to elimination of second set of CRTs as spares	Favorable	Unfavorable
2. Advantages due to familiarity by maintenance personnel of existing CRTs	Favorable	Unfavorable
3. Interchangeability of CRTs with other radars	Favorable	Unfavorable
4. Savings due to quantity purchase program for CRTs	Favorable	Unfavorable

For a prototype procurement such as the RLS, commonality does not play any significant role. However, the project engineer must be aware of this factor and must incorporate the maximum degree of commonality in any quantity procurement and describe how it is to be achieved in the technical proposal.

3.10 Cost Effectivity

Effectivity, as it relates to cost, is another new word that has arisen because of large Department of Defense procurements. The term relates to a concept that reflects the total cost to which a customer may be subjected if a particular design of equipment is purchased. An illustration of this point would pertain to the evaluation of two different designs of high-performance aircraft that are offered by two competitors. Assume that aircraft A costs $100,000 per unit more than aircraft B. Aircraft A can land and take off on the average 10,000-foot runway. However, although aircraft B is less costly, it requires a 15,000-foot runway. The procuring agency making a study of runway lengths finds that the cost of increasing the length of the airport runway far exceeds the total cost differential between aircraft A and aircraft B. The evaluation of the total cost picture constitutes cost effectiveness. In the case of the two aircraft designs, the more expensive aircraft A would be the design to purchase.

The concept of cost effectiveness should be considered by the project engineer in selecting a design approach for a procurement and should be discussed in detail in the proposal particularly when the technical proposal requirements call for a cost effectiveness discussion.

3.11 Delivery

An analysis of the selected design approach would have to be made by the project engineer to establish a schedule which would satisfy the deliv-

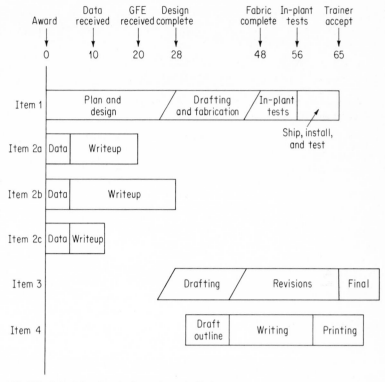

FIG. 3.5 Schedule of project events and effort.

ery schedule of the procurement. The establishment of the delivery schedule requires that the major milestones and the dates that each milestone must be completed in order to maintain the desired deliveries be identified. Factors such as work loads and the availability of facilities must be considered in working up each phase of the schedule. In addition, the support items called for in the contract must be considered and planned. Figure 3.5 depicts a Gantt Chart type of schedule plan that applies to the radar landmass procurement and that would be included in the proposal submitted by Acmen Electronics Company.

3.12 Summary

To summarize, the steps involved in dividing a system into logical subsystems using the RLS were illustrated. The tiers of design detail were presented with an explanation of which tiers would be of most interest to the project engineer. Block diagrams reflecting different tiers of information were presented.

A discussion of how the basic requirements, namely, specification, compatibility, competitive cost, and delivery, were considered by the project engineer in arriving at a fundamental decision for a design approach, using the FSS and optics systems as a case. The steps were described which lead to the conclusion that a single-FSS and optics system was the best design approach to be taken.

PROBLEMS

1. Company X has received a performance specification soliciting technical proposals for the design and fabrication of an elevator system for use in lifting emergency supplies from one deck to a deck above. The distance to be traversed is 22 feet. The maximum load to be carried is 500 pounds, and the elevator carrying area is to be 9 square feet. The system must be capable of withstanding 3-G shock and will be subjected to normal environmental hazards of shipboard application. Reliability of operation is of prime importance, and the elevator system must be designed with a backup system in the event of failure to the main activating system. The system must be able to operate on either 110 volts a-c or d-c and must be hydraulic, pneumatic, or mechanical in design. Timely delivery is also of foremost importance. Describe in detail the logic in arriving at a design approach to be proposed.

2. For the elevator system discussed in the answer to question 1, lay out block diagrams reflecting tier I and tier II design details.

Chapter Four

Preparation of the Proposal

4.1 Technical Proposal Requirements

The Acmen Electronics Company is one of the several companies judged by the procuring activity to be qualified to participate in the RLS Program. As noted in Section 2.1, the TPR is one of the documents that comprise the RFP package. The TPR specifies the sequence of material to be presented, the type of information required, the areas for which details of engineering hours, manufacturing hours, and material costs are required, and other information to be evaluated by the project team of the procuring organization. The technical proposal must usually be organized and presented in the following three sections or volumes:

1. *Engineering Presentation*
2. *Implementation Plan:* e.g., personnel, facilities, and scheduling.
3. *Logistic Support:* e.g., support documentation, spare parts, and installation.

Since the technical proposals will be used to establish which offerors will be considered technically qualified for the procurement, the project engineer's main concern will be to direct the preparation of the best possible documents. Some procurements will contain TPR which are very definitive. Other procurements may set forth the proposal requirements in very broad manner, thereby giving the offeror wide latitude in what may be described or presented. As a general rule, the extent of detail that is cited and the preciseness of the TPR depend on the desires of the customer's project engineer. Having a detailed concept of what is described and what the problem areas are, the customer's project engineer will be very specific as to the type of information required. The areas that the TPR identifies for discussion in the proposals usually include the following:

1. Areas which are critical to the successful design of the equipment: In the case of the RLS Program, such an area might be the design of the optics system, which is critical to meeting the resolution requirement of the specification.

2. Areas which by their nature pose as difficult engineering problems that must be successfully resolved: The design of the radar system which would simulate the characteristics of the AN/APQ-28 radar (accuracies, beam pattern, etc.) would be such an area to be covered in the engineering proposal.

For government programs, the procurement regulations require that the relative importance of each area of the TPR be indicated and that the general criteria for evaluation be provided for all offerors. It is therefore very important that the offeror's project engineer study the TPR document in conjunction with the specification and prepare the technical proposal precisely as required.

The proposal requirements for the engineering presentation (Volume I) portion of the RLS procurement will be discussed. The design information that is required was derived from the specification that was summarized in Section 2.4. The proposal requirements for Volume II (Plan for Implementation) and Volume III (Logistic Support) will not be treated in detail since the information of these volumes generally will be a secondary factor in the selection of the contractor. The following illustrates a TPR that would be used for the landmass simulator:

TECHNICAL PROPOSAL REQUIREMENTS
DEVICE 5A1

Radar Landmass Simulator
Volume I. Engineering Presentation
   The offerors shall describe the design they propose to use for meeting the requirements of Specification N1001 in the areas defined herewith. The Engineering Approach portion of the proposal shall be organized to present the following material in the order presented below:
   1. Introduction
      a. General Design Approach

2. Technical Discussion

  * *a. General Design.* Describe the general design of the simulator describing how the functions of the various subsystems will be performed and how they will interface with each other to provide the performance objectives of the specification. Supplement the narrative with block diagrams and schematics as appropriate.

  * *b. Resolution.* Present a summary of the engineering approach that will result in meeting the resolution requirements of the specification based on the areas to be simulated and the size limitations of the trainer. Provide calculations and diagrams to support the presentation.

  * *c. Simulated Radar System.* Provide a description of the design of the simulated radar indicating how the radar characteristics of range and bearing accuracies, beam configuration, pulse rate, etc., that are specified will be realistically simulated.

  d. Test Program. Describe the test program for components, subsystems, and complete system that will be used to verify the following:

    (1) Compliance with specification accuracies.
    (2) Overall performance capabilities.
    (3) Reliability requirements.
    (4) Environmental capabilities.

3. Hour and Material Breakdown

  a. For each major subsystem of the Radar Landmass Simulator, provide the breakdown of engineering hours, manufacturing hours, and material costs.

4. Schedule

  a. For the Radar Landmass Simulator, provide in a Gantt Chart form the scheduled effort for the following:

    (1) Design engineering
    (2) Drafting
    (3) Fabrication
    (4) Test and checkout
    (5) Government checkout

5. Experience and Facilities

  a. Provide a summary of the offeror's experience and facilities, including key engineering personnel for each of the design subsystems that comprise the simulator.

---

* Denoted as critical areas of equal weight. A proposal deficient in any critical area will be judged to be unacceptable for this Radar Landmass Simulator procurement.

The areas that the TPR identifies must be presented exactly as required in the technical proposal. The evaluation by the project team of the procuring activity will result in determining the relative excellence of each proposal.

It is incumbent upon the project engineer to concentrate on those areas deemed critical to the procuring activity. If such areas are not identified in the TPR, the project engineer may have to conjecture as to which of the areas called for are the more important areas to the procuring agency. For instance, if the Radar Landmass Simulator TPR did not identify the critical areas, an offeror would have to make the decision as to which technical areas to emphasize in the proposal or seek out such information through contacts and liaison effort. Often the offeror's marketing organization is used to obtain such types of intelligence.

In order to put all offerors on an equal footing as far as the relative importance of the areas to be evaluated are concerned, procurements by the government and many private concerns not only require that the TPR identify the critical or important areas but also require that the TPR indicate the relative importance of each area. The TPR for the RLS procurement presented above notes that each critical area is considered of equal weight.

4.2   Evaluation Factors

The project engineer for Acmen Electronics Company must be constantly sensitive to the thinking and desires of the project engineer of the procuring activity. Whereas this text is concerned with the role of the project engineer for the offeror and ultimately the contractor, consideration of some of the functions of the peronnel representing the procuring activity is in order.

One of the prime functions of the procuring activity's project engineer is to head up the evaluation team which is responsible for judging the most qualified offeror. The evaluation of each of the proposal areas that the TPR identifies must be equated with a common set of criteria which constitute the evaluation factors.

The evaluation factors are the criteria that the procuring project engineer uses for scoring the proposals that are received. Frequently, the factors are weighted. The specific factors and their scoring weights are kept in confidence, but astute offerors competing for a procurement often can deduce much of this important information which might give them an edge on other offerors.

For the RLS procurement, it might be assumed that the project engineer of Acmen Electronics, through discussions with the procuring activity engineers and through information obtained by the Acmen marketing force, had been able to obtain the following background information:

1. The radar landmass simulators procured in the past did not provide for shadow effects and were not able to achieve a resolution of better than 1,000 feet for a comparable area size. The procuring activity personnel during different preproposal briefing and discussions dwelt upon

these two areas and emphasized the importance of those particular requirements.

2. Previous radar landmass simulators were overly complex and difficult to maintain. The proposal evaluators have voiced concern regarding trainers in the field that were too complex to operate because of basically unsound design.

3. The operational radar equipment to be simulated operates with great accuracy. The simulator design must also be capable of achieving the same degree of accuracies in order to provide realistic training.

4. Previous devices used to train radar operators did not possess the flexibility for displaying new geographic areas nor were they able to be modified to incorporate added capabilities.

Based on the above type of intelligence the project engineer can deduct that the proposal evaluators will score each proposal heavily as to how well the points noted will be addressed. A reasonable conclusion would be that the major evaluation factors for each proposal will include the following points: (1) solution of critical and problem areas such as design to achieve shadow effects and required resolution, (2) soundness and simplicity of design, (3) design features for achieving required accuracies, and (4) design features which enhance the flexibility of the system.

The above points therefore should be constantly reviewed to assure that the content of the technical proposal emphasizes the areas that the procuring activity is very sensitive about.

4.3   Proposal Outline

Prior to initiating the proposal writeups, the project engineer should prepare an outline for the engineering approach portion of the proposal. The order in which the material is to be presented is generally established but the points to be discussed must be identified in each of the proposal sections.

The following is a typical proposal outline which illustrates the content of the engineering approach portion or volume for the landmass simulator.

ACMEN ELECTRONICS COMPANY
ENGINEERING APPROACH OUTLINE
FOR RADAR LANDMASS SIMULATOR

Volume I. Engineering Presentation
    I. Table of Contents
   II. List of Illustrations
  III. Introduction
      A. Company Technical Credentials
      B. Statement of Broad Design Concepts

   IV. Technical Discussions
      A. General Design (critical)
      B. Resolution (critical)
      C. Simulated Radar System (critical)
      D. Test Program
    V. Breakdown of engineering hours, manufacturing hours, and a breakout of subcontract items and their costs, material cost on basis of a fixed-price incentive contract
   VI. Schedule for all contract items including schedule of procured material and subcontracted items
  VII. Description of facilities availability for procurement
 VIII. Experience
      A. Personnel experience
      B. Summary of previous contracts for similar equipment

4.4 Proposal Language

The completion of the Engineering Approach Outline is the signal that the customer's requirements have been studied and analyzed and that the project engineer has established to sufficient degree of detail what will be offered. In addition, project engineers would, at this point, have established the very important requirement as to the scope of effort required for the procurement. It is their task now to express in simple understandable language what their company is offering and why it is to the advantage of the customer to procure the items from their company and not from one of the competing firms.

It should be mentioned at this time that the drafting of the proposal writeup is closely associated with the compiling of cost and time elements which will make up the proposed price and schedule. The project engineer, having established the scope of effort which is expressed as tier I or tier II of design detail, will farm out to different groups in the company's organization the task of accomplishing basic technical writeups and the time and cost estimates of materials, engineering, and manufacturing. The selection of the tier of detail to be reflected in the technical and cost proposal will be contingent on the customer's proposal requirements. At any rate, after the information is compiled, the project engineer will assemble the data and organize it into the proposal.

The narrative of the technical proposal should be written in a manner and style which will present a clear and concise picture of what is being offered and be sufficiently varied in style to hold the interest of the reviewer. A certain amount of psychology can be fruitfully exercised here. For instance, if it was determined that the customer's evaluating engineer was relatively inexperienced in optics, the proposal could discuss some of the basic fundamentals of the subject as applied to the proposed design, thereby enhancing the reviewer's grasp of the approach and facilitating an appreciation of the advantages that the particular design offers. One of the worst things that can happen is to submit a technical proposal that presents difficulties in understanding what is being offered. A reviewer cannot be expected to recommend an approach that cannnot be readily understood, regardless of its inherent merits.

4.5 Introduction of the Proposal

In the proposal introduction, the discussion must be confined to those qualifications that relate directly to the procurement. For instance, it might be an established fact that Acmen Electronics Company enjoys a reputation for being expert in the field of accelerometers, but since this particular capability is unrelated to the procurement under consideration, the impact of the proposal would be dissipated if the proposal discussed this capability of the company.

The following would illustrate some of the points that would be included in the introduction section for the Radar Landmass Simulator proposal:

A. Statement of Qualification
   1. Accomplishments of Acmen Electronics Company in developing and designing systems involving optics, photography and electronics.
   2. How the accomplishments cited above qualify the company as competent to design the various subsystems of the Radar Landmass Simulator.
B. Company Technical Credentials
   1. Tabulation of the government and commercial contracts and orders in areas related to those of the Landmass Simulator, giving a résumé of type of contract, delivery and cost performance in each case.
   2. Summary of company sponsored research programs and how they might relate to the procurement.
   3. Summary of any distinctive achievements especially if the area of achievement relates to some facet of the procurement under discussion.
   4. Review of technical problem areas that were encountered in the past and how the company, by creative engineering and ingenuity, was able to solve the problem.
   5. Review of successful programs that were experienced by users on past procurements because of sound and simple equipment designs.
C. Statement of Broad Design Concepts
   1. This part of the Introduction should serve as the summary of what is to be offered, describing the design approach proposed and how the design will meet the specification requirements.

4.6   Technical Discussion

The structure and arrangement of the proposal material must follow the sequence of material as called for in the TPR document for the RLS. The material is required as follows:

General Design (critical)

In this section of the proposal the project engineer would provide the design approach for each of the major subsystems of the simulator, remembering that the evaluator is attempting to seek the answer to one predominant question: "Will the proposed design result in a trainer that satisfies the specification requirements and how will the objective be achieved?"

To address that question, the project engineer should address each of the trainer subsystems, describing how each will interface with one another, how the specification requirements will be met, and how the deducted evaluation points will be satisfied. The subsystem of the simulator that would be addressed in the proposal would include the following: FSS and optics, transparency and its systems, detector, circuits for achieving radar effects, trainee display, and control system.

This section of the General Design discussion should also include a precise narrative describing what Acmen Electronics considers to be the major technical problems relating to the design and manufacture of the simulator and why they are considered to be problems. The discussion of the problem area can be very significant to the customer's evaluating engineer since it serves as evidence as to whether a particular offeror has an appreciation of the scope and complexity of the program.

For instance, the Shadow Generating Circuit of the RLS constitutes an area of development. A high point in a geographic area which is illustrated by a radar beam would throw a shadow on areas behind it in a very precise geometric pattern. The transparency of the simulator represents the terrain in two dimensions so that at some precise transparency point representing a high geographic area, the signal from the FSS must be blanked to create the synthetic shadow on the radar scope. If this very fundamental requirement is not recognized as a technical problem area by an offeror, then it would indicate that the offeror lacks an adequate grasp of the technical requirements and its complexity.

If major areas are treated with special consideration and creativity in the technical proposal, it would demonstrate to the evaluating engineer that an adequate study of the procurement had been made by the offeror and thereby generate a degree of confidence in the approach to be described. After the problem areas are described, it is mandatory that the technical proposal follow through and describe the proposed technical solution to each problem area.

In the case of the Shadow Generating Circuit, the design approach section of the proposal should clearly show how the problem areas would be solved in the simulator design. The technical proposal should describe the functions and basic designs of each of the elements of the block diagram of the major and significant subsystems of the simulator. The degree of detail of the description should be sufficient to permit the evaluating engineers to understand how the proposed system will work and to appreciate any of the advantages of the proposed system over those of competitive systems.

Since a sound and simple design approach is a feature to which the procuring activity proposal evaluators are sensitive, the general design section should expand the discussion on this feature.

For instance, if some company's design embodied a complex and expensive digital computer to generate a particular nonlinear function when a simple potentiometer in an analog computer could achieve the same result, then it can be judged that simplicity of design has not been achieved. An overly complex design will usually cost more and contain inherent complications of design, fabrication, and maintenance.

If an offeror conceives of a design approach that is unique, the proposal should describe in special detail how the unique design approach will meet the specification requirements and what specific advantages and benefits will accrue to the customer if the proposal is accepted. Quite often, a unique design will capture the fancy of the customer's project engineer and will result in influencing a favorable decision for contract award.

Another point judged to be a sensitive evaluation factor is the requirement for achieving the specified accuracies. Even though the specification may cite the overall accuracy requirements, the increments which contribute to the overall accuracy picture cannot be stated in the specification since they are functions of the detailed design approach which are undetermined at that time. In the case of the RLS, each of the following subdivisions must be designed to specified accuracies in order to maintain the overall accuracy of radar presentation which is stated as ± 5 percent in range as noted in the specification summary: (a) FSS sweep, (b) transparency details and registration, (c) optics light transmission, (d) detector sensitivity, and (e) display and controls.

In addition to the numerical accuracy values, there are other types of accuracies such as simulation fidelity and dimensional accuracies, all of which should be indicated in the actual proposal of the offeror.

As noted earlier, flexibility of design is considered an important feature in order to facilitate possible changes to the equipment design while still under contract. The customer is interested in design flexibility so that possible changes can be accomplished after the equipment has been delivered. Thus, it is to the interests of both the offeror or contractor and the customer to realize the maximum degree of design flexibility in the hardware.

To illustrate the concept of design flexibility, reference is made to Figures 3.4*a* and 3.4*b*, in which FSS and optical systems are shown. It would not be unreasonable to anticipate that a future requirement would involve using three transparencies instead of two. The third transparency would permit the storage of larger numbers of targets and greater terrain detail.

If a requirement incorporating the third transparency were adopted, the design depicted in Figure 3.4*a* could be modified by adding a third mirror and lens system and detectors. The FSS might have to be replaced by one which possesses greater light-generating capacity. The design shown in Figure 3.4*b*, however, would require not only a third lens and detector system, but a third FSS and redesigned synchronizers, feedback systems, and intensity regulators. Thus the design of Figure 3.4*a* is greatly superior to Figure 3.4*b* as far as flexibility of design is concerned.

Quite often, the requirements for a particular piece of equipment are not established in detail at the time of procurement. However, areas of potential change are generally known. The astute project engineer will make every effort to discover the questionable areas and plan the equipment design so as to facilitate possible future changes. Again, the extremely important role that effective liaison with the procuring party plays in a program is demonstrated since the information relative to possible future modification would be obtained from customer sources. This information coupled with an intimate knowledge of the equipment to be purchased will serve to guide the project engineer in planning and describing how the flexibility requirement of the proposed simulation will be achieved.

Resolution (critical)

For this area, the project engineer must present a description of how the resolution requirements will be met and provide the basic calculations to substantiate that the resolution requirements will be met. The text would address the following major points that determine the resolution capabilities of the trainer:

1. Transparency scale and method of recording terrain details
2. FSS and optics systems establishing how spot size will be achieved
3. Detector system and displays developing rationale supporting how resolution of 250 feet will be met

Simulated Radar System (critical)

This portion will present the engineering approach which will support how the specified radar characteristics will be simulated. For instance, it is not sufficient to state that the pulse repetition rate or the radar beam width of the AN/APQ-28 radar will be realistically simulated. The proposal must present the description of the design approach which supports the offeror's claim that the particular requirements will be satisfied.

Test Program

The test program for components, subsystems, and integrated systems for the proposed design, emphasizing the following, should be presented in this section of the proposal.

1. FSS and optics system facilities, test equipment, and procedures
2. Transparency carriage and registration (positioning) capabilities
3. Electronics systems
4. Description of quality controls and component selection procedures for meeting reliability specifications
5. Description of environmental test facilities and procedures

TABLE 4.1   Technical Evaluation Grid: Radar Landmass Simulator

Areas of evaluation	Weight	XYZ Corporation Score	Acmen Electronics Score	ABC Company Score
General Design:				
Subsystem	5	3	4	4
Interfaces	5	3	5	4
Shadow Effects	10	6	9	7
Simplicity	10	7	8	6
Flexibility	10	7	9	8
Total	40	26	35	29
Resolution:				
Engineering	20	15	18	16
Calculations	10	7	9	7
Total	30	22	27	23
Simulated Radar:				
Radar Beam	3	2	3	2
Accuracies	10	7	8	8
Display	4	2	3	3
Controls	3	2	3	2
Total	20	13	19	15
Test Program:				
System	4	2	3	2
Subsystem	2	1	1	1
Reliability	2	1	2	2
Environment	2	1	1	1
Total	10	5	7	6
Total Technical Score	100	66	88	73

In order to cover all the points of design and at the same time have the technical proposal arranged in accordance with the TPR, a clearly expressed cross-reference is essential, and in some cases redundancy may be in order. The project engineer should keep in mind that the review of the proposal may be done by several individuals, each assigned to review a certain section of the proposal. It can be assumed, for example, that an individual assigned to analyze only the section dealing with environmental testing of the simulator will read only that assigned section, even though a particular proposal may cover this particular point in another section. Thus, as a matter of self-protection, the project engineer must either repeat the discussion on environmental testing in all the applicable sections or make clear references to the section which covers the material.

Table 4.1 illustrates an evaluation grid form which is typical of that used by reviewing project engineers to evaluate proposals. The grid form summarizes the evaluation factors described above for each of the major areas of the RLS. Upon completion of the technical evaluation of each proposal, the customer's project engineer will total the evaluation points to obtain the relative technical standings. The table demonstrates how an arbitrary technical score of 88 for Acmen Electronics Company, 73 for ABC Company, and 66 for XYZ Corporation might be evolved upon completion of the proposal evaluations.

4.7   Material and Hourly Effort Breakdown

The cost breakdown of material, engineering hours, and manufacturing hours is usually required for each deliverable contract item, and a specific form for use in presenting the cost information is generally provided. Table 4.2 illustrates such a form for cost breakdown presentation. Procedures for estimating the number of hours required for a particular project and the various factors that enter into the cost estimate will be discussed in detail in Chapter 6. Suffice to say, the project engineer must segregate the types of effort required for a procurement, e.g., electrical, engineering and mechanical engineering, assign the estimating chore to the cognizant specialist, assemble and organize all the different cost inputs, modify the figures as necessary, and present them in the form required by the technical proposal.

Very often, the procuring agency will require a more detailed breakdown of costs than that reflecting each deliverable item as discussed. In the case of the simulator under procurement, the TPR stipulate that the cost breakdown of the simulator (contract item 1) reflect the costs of the four major systems, e.g., the input, detection, signal processing, and display systems. To present the detailed cost breakdown of the simulator, a chart such as shown in Table 4.3 would be used. The figures in the

TABLE 4.2   Itemized Cost Breakdown: Radar Landmass Simulator

Elements	Item no. 1		Item no. 2	
	Hours	Dollars	Hours	Dollars
Material	..	XXX	..	XXX
Material handling %	..	XXX	..	XXX
Subcontracts	..	XXX	..	XXX
Subcontract overhead %	..	XXX	..	XXX
Engineering	XX	XXX	XX	XXX
Engineering overhead %	..	XXX	..	XXX
Manufacturing	XX	XXX	XX	XXX
Manufacturing overhead	..	XXX	..	XXX
Other cost factors	..	XXX	..	XXX
Subtotal	..	XXX	..	XXX
G&A rate, %	..	XXX	..	XXX
Subtotal	..	XXX	..	XXX
Profit, %	..	XXX	..	XXX
Total cost	XX	XXX	XX	XXX

lower right-hand box would be the total of dollars and hours for the item.

4.8   Scheduling

In the past, the offeror usually made guesses of the time that would be required for completing various tasks. Usually the guesses were based on past experience on related or similar procurements, but all too often an offeror made educated guesses based largely on what they felt the customer wanted to hear. Because of many sad experiences where deliveries were months or years late, procuring companies and government agencies are requiring that proposed delivery schedules be based on and substantiated by a PERT (Program Evaluation Review Technique) analysis.

An experienced company can present a program schedule based on estimates and experience which are accurate and sound. Such estimates must, of course, be realistic. If, for example, the assembly of the readout system requires as estimated 50 man-weeks and the facilities and qualified personnel necessitate a minimum of 6 weeks to perform the task, the offeror cannot, in good conscience, indicate a schedule of less than 6 weeks for the task.

The reviewing project engineer of the procurement agency can be expected to challenge any schedule, and if the periods of effort cannot be logically defended by the offeror, the success of obtaining the contract award can be jeopardized.

TABLE 4.3 Detailed Cost Breakdown of Simulator (Item 2)

Elements	Input system Hours	Input system Dollars	Detection system Hours	Detection system Dollars	Signal processing Hours	Signal processing Dollars	Display system Hours	Display system Dollars	Total Hours	Total Dollars
Material		XXX		XXX		XXX		XXX		XXX
Material handling %		XXX		XXX		XXX		XXX		XXX
Subcontracts		XXX		XXX		XXX		XXX		XXX
Subcontract overhead %		XXX		XXX		XXX		XXX		XXX
Engineering	XX	XXX	XX	XXX	XX	XXX	XX	XXX	XX	XXX
Engineering overhead %		XXX		XXX		XXX		XXX		XXX
Manufacturing	XX	XXX	XX	XXX	XX	XXX	XX	XXX	XX	XXX
Manufacturing overhead %		XXX		XXX		XXX		XXX		XXX
Other cost factors		XXX		XXX		XXX		XXX		XXX
Subtotal		XXX		XXX		XXX		XXX		XXX
G&A rate, %		XXX		XXX		XXX		XXX		XXX
Subtotal		XXX		XXX		XXX		XXX		XXX
Profit, %		XXX		XXX		XXX		XXX		XXX
Total cost		XXX		XXX		XXX		XXX		XXX

The TPR for the landmass simulator require that the scheduled effort for item 1 be presented for design, drafting, and test in a Gantt Chart. Figure 4.1 illustrates how this particular information can be presented in such a chart.

The overall schedule for the different contract items is shown in Figure 3.5, Schedule of Events in Program, which can be adapted to meet the terms of the TPR.

Because of the importance of delivery of the items required in a procurement, the offeror must be prepared and able to expend the required effort at a rate which will permit maintaining the delivery schedule. Quite often, this may require hiring additional personnel, procuring required capital equipment, subcontracting, or undertaking other special steps. The plans of action that would be required should be described in sufficient detail to substantiate how the procurement schedule might be met.

If offerors are in a position such that they can improve the delivery schedule, they should emphasize this fact in their proposal and document the basis for the improved delivery schedule. Invariably, the company that can offer an improved delivery schedule has the know-how and experience for the procurement and its technical and cost proposal should reflect its advantages.

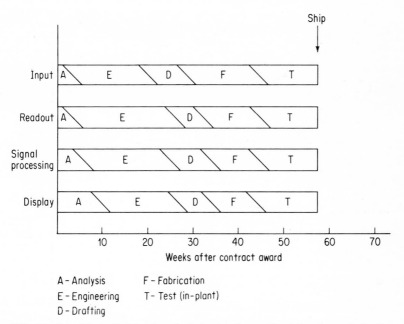

FIG. 4.1   Scheduled effort for item 1, Radar Landmass Simulator: A—analysis; D— drafting; E—engineering; F—fabrication; T—test.

4.9 Facilities and Personnel

Of particular importance is the availability of the offeror's facilities and personnel for meeting the schedule of a program. One frequent cause of difficulties in meeting contract schedules is lack of availability of personnel or facilities due to other projects. The reviewing procuring project engineer will be interested in determining the overall schedule of offeror's projects and their ability to handle the procurement under negotiation.

Figure 4.2 graphically portrays the engineering capacity of the Acmen Electronics Company. The figures for engineering hours are the totals of electronic, mechanical, and other engineering disciplines. The solid line represents the engineering loading that is and will be required to perform under existing contracts. The dotted line represents the additional number of engineers that would be required if the company were to receive the award for the contract on January 1.

When estimates of projected effort are made, based on a weekly interval, the number of hours for a person is given as 40 (40 hours per workweek). There are numerous ways to show the information, and in the case of Figure 4.2, the numbers on the left side of the chart would be multiplied by 40 to indicate the number of engineering man-hours available for any particular month.

Figure 4.3 gives the same type of information discussed above, except that the man-hours relate to manufacturing.

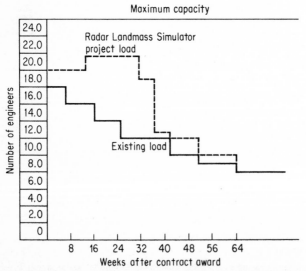

FIG. 4.2 Schedule of Acmen Electronics Company engineering load.

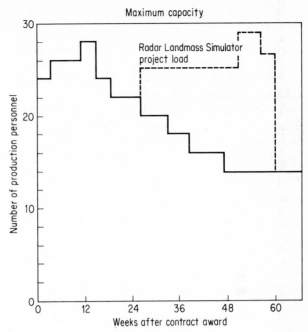

FIG. 4.3   Schedule of Acmen Electronics Company manufacturing load.

The information shown in Figures 4.2 and 4.3 serves as a very useful tool to permit the reviewing engineer to see at a glance the ability of a company to handle a particular project as far as personnel availability is concerned. The text of the proposal should expand on the company loading to demonstrate that the personnel requiring special engineering disciplines experience are available to properly execute the requirements of the project under consideration.

4.10   Description of Facilities and Experience

The description of the facilities should include a schematic layout of the overall plant floor space, a list of machine tools and other capital equipment, and a description of laboratory facilities and other physical property of the company. In particular, the proposal should describe in detail those facilities which are necessary and which will be used to complete the project under consideration. The general theme that has been expressed previously may be reemphasized: Offerors should stress any point which will contribute to placing themselves in a more favorable competitive position.

The portion relating the experience of Acmen Electronics Company should treat two areas:

1. The experience of the company in completing contracts involving the same type of equipment. This portion should list each contract so cited and briefly describe how the effort and product are similar to the equipment under procurement.

2. Experience of key personnel who will be assigned to the project. This portion should include a brief description of what each individual's function will be in the program, this person's title, and a description of the background, experience, and qualifications of the individual which are necessary to carry out these functions efficiently.

4.11 Summary

The TPR is one of several documents that comprise the RFP that is distributed to qualified companies on a procurement solicitation. A TPR for a major procurement generally requires each offeror to submit a proposal in three volumes: engineering presentation, implementation plan, and logistics support.

The engineering presentation volume of a technical proposal is the prime document which determines whether the design an offeror proposes is acceptable or unacceptable.

The type of information that is generally requested to be included in the engineering approach volume are discussions of technical areas that are considered to be critical to the equipments function or areas that represent difficult engineering problems. For the RLS, the areas that were to be addressed were: (a) General Design, (b) Resolution, (c) Simulated Radar System, and (d) Test Program.

The procuring agency or organization must evaluate each technical proposal against a predetermined set of criteria to establish which company offers the best technical design for the equipment under procurement. The project engineer's sensitivity to the identity of the customer's evaluation criteria would permit the structuring of the technical proposal to satisfy the criteria, thereby enhancing the chances for contract award.

The TPR set forth the organization and type of material that is to be presented in the offeror's technical proposal. The type of information usually includes the technical description, proposed schedule, cost breakdown, experience, and description of facilities. The technical description should be written in such a way as to cover thoroughly the evaluation factors that will be used, the identity of which the offeror might learn or deduce from previous similar procurements. In addition to the probable evaluation factors, the project engineer must be certain that the technical description covers all the significant specification requirements as well as the points in the TPR document.

The cost breakdown involves separating the basic cost elements of material, engineering, and manufacturing that constitute the bid price. The degree of detail of the cost breakdown will be specified in the TPR document, and it is important that the project engineer understand the requirements and comply with them.

The cost breakdown should be presented in a chart form or on forms provided by the customer. The schedule of required effort should be presented in a Gantt Chart and should clearly show the sequence of effort that is proposed for the project. In addition, the overall schedule of the total business that offerors have, including the additional effort that the proposed project would entail, should be presented. This information should be shown in relation to the total maximum capacity that offerors possess for their personnel and facilities.

The description of the offeror's facilities and experience should emphasize those facets and disciplines that are directly related to the proposed procurement, and should show how these company resources will be applied to the proposed project.

Briefly stated, the technical proposal should cover all stipulated and anticipated areas on which the proposals will be evaluated, and the project engineers must emphasize every point that will contribute to fostering a competitive advantage for their company.

The procuring activity identifies in the TPR document the information to be presented in proposals. The proposals are evaluated to establish the most highly qualified offeror. The evaluation factors established by the customer are used as criteria for the scoring of proposals and are often made known to the offerors. The project engineer in charge of preparing a technical proposal should prepare an outline and assure that all points of TPR are fully addressed.

In addition to the technical discussion, the proposal usually requires a detailed breakdown of material and labor costs and overhead rates. Also, the schedule of effort proposed for meeting the delivery requirements of the equipment is required in each proposal.

To facilitate the evaluators' reading and understanding of the proposals, the material should be organized logically and written in the most lucid manner possible. Special attention must be given to the TPR to assure that all the required information is provided in the form and sequence specified.

PROBLEMS

A company has received an RFP for a special navigation computer for use in aircraft. A functionally identical computer of a mechanical design is in use in large aircraft. The performance specification requires that the computer under consideration occupy one-half the volume of its present counterpart but have the same reliability and identical accuracies, meet equivalent environmental tests, and in essence must be functionally identical. List the points of discussion with a short explanation relating to the following areas of technical proposal: (1) understanding the problem, (2) soundness of design approach, and (3) ease of maintenance.

Chapter Five

Contract Schedule

5.1 Description

A RFP on any procurement will include a proposed contract schedule which forms the basis for the final contract. A contract schedule is generally divided into sections, each setting forth the specific language and terms which establish the contract requirements.

The words contract schedule might mislead an individual into concluding that the document applies only to times of delivery. The references to the calendar are only one part of the contract schedule.

Far more significant from the contract point of view, and therefore to project engineers and their companies, are the contract clauses, terms, and other conditions which are in the schedule. Usually any issue of a dispute between the two parties in a contract revolves around the application or interpretation of some element or phrase of the schedule language.

The contract schedule is the most important legal document of a procurement and takes precedence over any other document in the event of any conflict of language. Usually, the order or precedence of documents relating to a procurement is cited in one of the sections of the schedule in order to preclude any misunderstanding in conflicting areas.

To return to the example of the procurement of the RLS, the following is some typical schedule language which will be analyzed and discussed to illustrate the various principles involved:

SECTION A

Item	Contract schedule	Quantity
1	Radar Landmass Simulator, Device 5A1 delivered and installed at site	1 each
2	Engineering Design Reports (Government shall be allowed 30 days for review and comments or approval on all reports)	
	(*a*) Terrain Readout System	2 copies
	(*b*) Signal Processing System	2 copies
	(*c*) Display System	2 copies
3	Engineering Drawings	2 sets
4	Installation and Maintenance Manuals	2 sets

SECTION B   *Delivery*

The following items to be furnished under Section A above, shall be furnished in accordance with the following schedule at the destinations noted below:

Item 1: One unit shall be delivered and installed by the contractor and made ready for acceptance 62 weeks after date of contract at the Naval Training Station, Orlando, Florida, Building 7852. The government will take three (3) weeks for performing acceptance tests. The contractor is responsible for correcting any deficiencies identified during acceptance testing and is responsible for any delays in trainer acceptance due to the deficiencies.

The following items shall be delivered to the Contracting Officer, Naval Training Equipment Center, Orlando, Florida, in the quantities and during the periods listed below:

Item 2a: Shall be delivered 20 weeks after date of contract.
Item 2b: Shall be delivered 25 weeks after date of contract.
Item 2c: Shall be delivered 12 weeks after date of contract.
Item 3: Shall be delivered 73 weeks after date of contract.
Item 4: Shall be delivered 70 weeks after date of contract.

SECTION C   *Description of Item*

Item 1 shall be in accordance with Performance Specification 1001, Radar Landmass Simulator, Device 5A1, and all amendments thereto.

Item 2 shall be in accordance with U.S. Naval Training Equipment Center Specification 1001, Standards for Engineering Design Reports.

Item 3 shall be in accordance with U.S. Naval Training Equipment Center Specification 1001, Standards for Engineering Drawings.

Item 4 shall be in accordance with U.S. Naval Training Equipment Center Specification 1001, Requirements for Maintenance Manuals.

SECTION D *Government Furnished Property*

The following items will be furnished to the contractor by the government in the quantities and at the times indicated below:

Item	Description	Quantity	Delivery to con-tractor from date of contract
1	AN/APQ-28 Radar Display Unit Part APQ-28-64	2	20 weeks
2	AN/APQ-28 Radar Control Box	1	20 weeks
3	AN/APQ-28 Synchronizer, Part APQ-29-9	1	20 weeks
4	Design Data on AN/APQ-28 Radar System	Lot	10 weeks

SECTION E *Design Approval*

The formal approval of the Engineering Design Reports required in item 2 shall constitute authorization for the contractor to proceed with the detail design, manufacturing, and other effort necessary to meet the requirements and schedule of the contract. The government shall approve or reject, with comments, any submission within thirty days. In cases of rejection, the contractor shall submit a revised report within thirty days after notification of rejection.

SECTION F *Precedence of Documents*

In the event of any conflict, the following is the order of precedence of documents relating to this procurement:
  (1) Contract Schedule
  (2) Referenced Standard Contract Clauses
  (3) Performance Specification No. 1001 for Device 5A1
  (4) Referenced Specifications of Performance Specification
  (5) Contractor's Proposal

SECTION G *Inspection and Acceptance*

Item 1 shall be set up and dynamically activated at the contractor's plant for preliminary inspection prior to shipment to the site destination. Preliminary inspections shall be made by the Contracting Officer's representative to verify that the performance characteristics of the Radar Landmass Simulator are in accordance with the contract requirements and shall require no more than three days if no major deficiencies are observed. The government reserves the right to direct the correction of deficiencies and conduct subsequent in-plant inspections of the unit prior to the trainer being shipped. Contractual acceptance of item 1 will be made by the Contracting Officer's representative after delivery and installation is satisfactorily completed. Acceptance tests will be made in accordance with the Acceptance Test Procedures Report. However, the government reserves the right to perform any additional tests which are considered appropriate to verify the performance characteristics of the simulator and direct corrective action by the contractor if deemed necessary.

SECTION H   *Liquidated Damages*

In case the contractor fails to deliver item 1 on or before the date specified in Section B of this contract, the contractor shall pay to the government as liquidated damages the sum of $100 for each day of delay up to a maximum of 90 calendar days and the sum of $50 per day for the succeeding 90 days.

SECTION I   *Site Availability*

The government will have an air conditioned room 20 by 30 feet in size, 8-foot ceiling supplied with 100 KVA, 120/208, 3 phase, 4 wire, 60 Hertz power available to house and operate the Radar Landmass Simulator described in item 1, at the Naval Training Center in Orlando, Florida, no later than ten (10) months after date of contract award.

5.2   Analysis of Contract Schedule

The contract schedule in conjunction with the specification and referenced documents establishes what is required, the delivery dates, quantities, and conditions relating to the procurement. It is essential that the project engineer make a complete and detailed analysis of the various sections of the schedule, identify potential problem areas, and evaluate the impact of any contractual requirement on the company's ability to satisfy the contract and make a profit.

In analyzing the various sections of the schedule, the following questions must constantly be kept visible to all members of the project team:

1. What is required?
2. When are they required?
3. Will the company have the necessary resources and facilities to produce what is required on schedule?
4. What contractual conditions may impose special demand or problems on the company?
5. What contractual obligation does the customer assume?

Section A of the schedule lists the various "line" items of the proposed contract and the quantities required. Items 2, 3, and 4 are commonly referred to as support items and the amount and type of effort required to produce and deliver such items usually is significant. There have been numerous cases when project engineers, in their zeal to concentrate on the primary hardware items, gave the support items secondary attention with the result that costs to produce the support items proved inadequate. Thus an otherwise profitable contract turned out to be a loss to the company.

The delivery schedule of the contract items is contained in Section B. It should be kept in mind that the delivery requirements specified by the procuring agency or customer have been carefully thought out and are rarely unreasonable. Usually the schedule is based on actual performance by some company for similar or related items. Project engineers can assume with a great deal of confidence that even if they feel that a particular delivery schedule is unreasonable (for their company) there are always competitors who may not share these sentiments.

However, since it has been previously established that the project engineer's company, Acmen Electronics, is well qualified for this procurement, the delivery schedule can be established as being reasonable. The project engineer will be required to portray the project schedule graphically during discussions with management as well as subordinate personnel. Figure 3.5, Schedule Events in Program, is comparable in scope to the tier I block diagram and illustrates a good example of the type of schedule Gantt Chart that would serve the project engineer's requirements.

In deriving the Gantt Chart of Figure 3.5, the project engineer must divide each contract item into segments of effort. Since the delivery date of each item is inflexible, the best approach is to start from the delivery dates and work backward to the contract award date.

The segments of major types of effort during the 65 weeks required for delivery of the simulator of item 1 are as indicated in Table 5.1. The schedule of phases starts with the project objective which is trainer acceptance and works backward to time zero which is the date of contract award.

In like manner, the reports (item 2a, 2b, and 2c), drawings (item 3), and manuals (item 4) all can be shown on the Gantt Chart with the

TABLE 5.1 Phase Schedule—Radar Landmass Simulator

Phase of effort	Period of effort	Phase complete weeks from award
Customer Acceptance	3 weeks	65
Install and Test	4 weeks	62
Pack and Ship	2 weeks	58
In-Plant Tests	8 weeks	56
Fabrication	12 weeks	48
Drafting	8 weeks	36
Design	22 weeks	28
Planning and Data	6 weeks	6
Contract Award	0	0

segments of effort identified as shown in Figure 3.5. It should be noted that aside from general planning and setting up of personnel assignments, active expenditures of effort do not take place until an appropriate time after the contract award date. Active effort on the drawing, for instance, cannot be initiated until sufficient progress has been made in the design effort to enable documentation of the drawings.

In addition to the schedule of major effort required for the program, the top of Figure 3.5 shows when the major milestones must be completed in order to maintain the project schedule.

Of particular significance is the fact that the customer (in this case, the government) is contractually obligated to deliver data and equipment by specific dates. The obligation of the government to deliver acceptable equipment and the government's responsibility for the support of such equipment are set forth in the Armed Service Procurement Regulations (ASPR) and other similar documents. It is vital that the project engineer know precisely the responsibilities of the government in this area. If the customer is some agency or company for which the ASPR does not apply, then the responsibilities of the customer must be specifically set forth in the contract.

Section C of the contract schedule, Description of Items, identifies each of the deliverable items to be procured.

Item 1, which is the RLS, has been described and analyzed in Chapters 2 and 3.

The requirements for each of the other items are called out in the specifications and documents cited for each item in question. Specification 1001 specifies the content of the reports, their arrangement, format, and other details of a similar nature. The specifications for the drawings and the one for the manual also specify similar format requirements. Since these specifications are straightforward and deal primarily with format, arrangement, and similar details, there is little requirement for the exercise of judgment and decision. The project engineer must be certain that the individuals assigned to the reports, manuals, and drawings are intimately familiar with the requirements that are applicable. Failure to comply with the specification details will invariably result in rejection of the items, thereby requiring expensive and time-consuming resubmissions. There are many case histories of companies who experienced losses on contracts because they failed to give proper attention to the detailed requirements of the side items such as manuals and drawings.

Section D, Government Furnished Property (GFP), describes what is to be furnished by the government for use on the contract. The government has a very real responsibility to provide items completely checked out and operating in a fully satisfactory manner by the dates indicated. Contractors assume the responsibility for normal diligence and care in handling and maintenance of furnished equipment, but they are not responsible for major equipment failures which are beyond their control.

Thus, the government's responsibilities which relate to meeting delivery dates, providing acceptable equipment, and providing adequate and timely support in the event of equipment breakdown are highly important as far as contractor performance is concerned. Many disputes involving large sums of money revolve about the following point: How valid is the contractor's claim that the government's (or customer's) failure to meet its obligations in providing Government or Customer Furnished Property caused the cost overrun and/or slippage in delivery? The issues in such claims are rarely clean-cut, and contractors will invariably be awarded consideration if they can prove that customers had failed to meet even a part of their responsibilities.

Section E, Design Approval, concerns the approval of the various reports for different systems. It should be noted that according to Section A the customer is under obligation to review and submit comments or approval within 30 days after receipt of each report from the contractor. The dates of approval are extremely important to the contractor since they establish when final detail designs can be completed and when drafting, production, and other similar tasks of the program can be initiated.

If for any reason the customer fails to submit its approval or comments on a report as required, the contractor has a legitimate claim for schedule slippage or cost increase. On the other hand, it is incumbent on the contractor to submit reports that comply with the requirements of the report specification cited in Section C. Usually specifications for items such as reports are very broad in scope and are subject to interpretation. To avoid any conflicts, it is essential that the project engineer maintain close liaison with the customer's project engineer in order to reach a complete meeting of the minds as to what is required. If a report is disapproved because of failure to meet any of the specification requirements, the contractor must resubmit a revised report as soon as possible and seek means to make up for any lost time due to the delays in obtaining report approval.

In any contract requiring research and development, it would be impossible to list every element and facet of the items to be delivered. Because of this fact, many areas are open to interpretation. It is essential that good faith and integrity be exhibited by both the contractor and the customer. If the contractor sincerely tries to live up to the requirements of the contract, usually the customer and the customer's project engineer are willing to interpret the specification requirements to the advantage of the customer if by so doing, the end product would not be compromised in any way.

Section F, Precedence of Documents, sets forth which of the contract documents would take precedence in the event of any conflict in terms of language between or among two or more documents. The precedence sequence noted is typical for a negotiated procurement that is the basis for the RLS case. One advantage of having the contract schedule carry the highest precedence is that a particular requirement feature of the proposed contract item may be changed by mutual consent during negotiations. The revision, which would differ from the specification and/or proposal, would be described in a special section of the contract schedule and would therefore supercede the specification and proposal language.

Different types of contracts might dictate variations from the precedence of documents. For instance, by definition a two-step procurement involves the incorporation of the contractor's proposal as the contract document instead of the specification. Therefore, the proposal might be designated as the contract document with the higher precedence.

The precedence of material that is established in a contract can be vital to either of the contracting parties. If, for instance, an offeror proposes a feature that is not consistent with the specification requirement in a particular area and the contract award is made without rectifying the inconsistency, the customer will be contractually bound to furnish something which was never intended. Because the contract cited the specification as having precedence over the offeror's proposal, the offeror (who is now the contractor) may have to provide a feature more expensive than what was proposed. It is important that the schedule, which is the document having the higher precedence, address any inconsistencies between the specification and the proposal submitted by the contractor so that both parties recognize exactly what is called for in the contract.

Section G, titled Inspection and Acceptance, establishes the requirements relating to the testing of the primary procurement item, namely, the RLS. The established ground rules for the inspections and acceptance provide protection for the contractor as well as the customer. The contract section establishes the criteria for acceptability and the periods during which tests are to be performed. Since the contractor does not receive final payment until after the articles being furnished are successfully tested and judged acceptable, it is essential that all aspects of the inspection criteria and periods be fully described in the contract in order to avoid controversy and unnecessary expenditures due to misunderstandings between the customer and the contractor.

Section H, Liquidated Damages, is one type of clause that contractors prefer to avoid. A liquidated damage clause in a contract forces contractors into a position wherein their failure to satisfy the basic performance and delivery requirement would cause them to suffer specific monetary penalties. A customer may impose a liquidated damages based on delivery when the failure of a contractor to meet a specific delivery date will result in the customer's suffering a hardship or loss. Project engineers, prior to permitting their company to become party to a contract, must first assure themselves that precautions against all the risks and contingencies that can be foreseen are taken by their company.

Section I, Site Availability, imposes on the customer the contractual obligation to have the housing and other facilities available by a specific date to permit the contractor to continue with the work under contract. Failure of the customer to meet these responsibilities under this section would probably work a hardship and subject the contractor to an expense which would be a valid basis for a claim by the contractor.

The various types of contract sections discussed above typify the content that might be contained in a contract. Since each procurement, whether by a government agency or a commercial firm, is unique, the content and terms of each contract would be tailored to the procurement and would be different. It is incumbent upon the project engineer to analyze all sections of the proposed contract schedule to assure that no requirement is overlooked. Failure to recognize the possible impact of a contract clause, such as liquidated damages, could very easily be the cause of severe losses on the contract.

5.3 Summary

A proposed contract schedule usually accompanies RFPs on major procurements and eventually is incorporated as a contract document. The purpose of the contract schedule is to identify precisely what is to be procured, how many, and when, and sets forth all the terms and conditions of a procurement contract. The contract schedule represents the most important legal document of a procurement and generally takes precedence over all other documents such as the specification. Most contractual and legal disputes on a procurement revolve around the interpretation and wording of the contract schedule.

PROBLEMS

For the procurement described in Section 5.1, the Government Furnished Property, items 1 and 2 listed in Section D, was received 40 weeks after contract award. Because of the delay in receipt of the items, the design and production effort were delayed 10 weeks.

1. Prepare an outline to be the basis of a claim on the customer for a new delivery schedule and increased cost, citing how various areas of effort are affected.

2. Prepare a revised schedule graph.

Chapter Six

Estimating of Costs

6.1 Introduction

Most companies existing in today's dynamic economy derive significant portions of, and in many cases, all their business from competitive procurements based on bidding. Thus, the growth, welfare, and frequently the very existence of an organization depend on its success in accurately estimating the cost to perform satisfactorily under different contracts.

The importance of accurate estimates is particularly evident in procurements which involve engineering design and which are based on some form of fixed-price contract arrangement. A new design involves a degree of uncertainty in terms of man-hours or other expenditures that might be required. Adequate provisions should be made in the cost estimate to cover those areas of risk. A company which does not provide for the risks which invariably occur in new designs may be successful in obtaining contract awards because of its low bid, but they will generally experience losses on their contracts. Obviously, no industrial organization can survive long when operating at a loss. On the other hand, a company that is too conservative in its cost estimate will not receive contract awards and will wither on the industrial vine.

In estimating the cost to be quoted on a procurement, project engineers must use all their experience, business acumen, and knowledge to derive an estimate which will not only enable them to get a contract award but will result in a reasonable profit for their company.

The difficulty in deriving accurate cost estimates is directly proportional to the amount of engineering design or development work that must be performed on a procurement. The first all-important requirement is that the project engineer, to the greatest possible extent, know what is required in a procurement and what design approach will be taken to fulfill the requirement.

6.2   Cost-estimating Concepts

The two basic categories of costs that must be considered are recurring and nonrecurring. An example of a nonrecurring cost would be design engineering, developmental or breadboard testing. Such costs generally are absorbed by the prototype units of the contracted equipment. A typical recurring cost would be the manufacturing cost for a system. A recurring cost is inherent in every unit of the equipment being produced.

Some of the factors that comprise the cost estimate for a production run of an article include cost of materials, manufacturing and assembly costs, quality-assurance procedures, special tooling, overheads, etc. In addition, the cost estimates must be modified to reflect anticipated rejection rates, learning curve impact, and other contingencies.

The cost estimate for equipment being produced from already completed engineering and development effort is basically a statistical exercise using current figures for material and labor rates and deriving the scope of manufacturing, assembly, and other types of effort from historical data of the same or similar equipment. A company that can implement some labor-saving technique or money-saving manufacturing procedure would enjoy an edge over competitors for a production contract.

The techniques and thinking process required to estimate the cost of designing, developing, building, and testing a prototype device or units of unique articles differ greatly from the disciplines previously mentioned for estimating production runs. Before the project engineer can begin to converge on a cost estimate, the device that is to be cost estimated must be defined in detail. The initial steps would require answers to the following type of questions relating to the equipment in question:

1. What is it? (Derived from the specification)

2. How will it work? (Derived from the system design approach that is selected)

3. What are its subsystems? (Derived from the block diagrams that are created)

4. What is the technical description of each subsystem? (This is derived from each of the elements of the block diagram.)

Once project engineers have clearly established the answers to the above questions, they would draw upon their judgment to render estimates of engineering and development hours, manufacturing hours, and material costs required for each subsystem. Prior to applying hourly rates, overhead, and other necessary cost factors, the estimate of hours is reviewed to eliminate such things as redundant cost elements.

6.3 Pitfalls of Estimating

The two major pitfalls of accurate estimating of costs for prototype equipment are errors in applying the mechanics of estimating and judgment errors.

In deriving the cost estimate, the project engineer should review the work to be sure that none of the following errors were made:

1. *Omissions:* Was any significant cost element forgotten? For instance, are there any bench checks planned and does the estimate include the engineering, material, and other costs for such effort?

2. *In accuracy of work breakdown:* Does the work breakdown adequately account for all the subsystems and efforts required of the device?

3. *Misinterpretation of the equipment data or function:* Is the interpretation of the complexity of the device accurate? Interpretations leading to excesses of complexity or simplicity will result in estimates that are either too high or too low.

4. *Use of wrong estimating techniques:* The correct estimating techniques must be applied to the device in question. For instance, the use of cost statistics derived from production runs of a similar subsystem and using such figures for a prototype device which requires engineering and/or development will invariably lead to excessively low estimates.

5. *Failure to identify and concentrate on major cost elements:* It has been statistically established that for any piece of equipment, 20 percent of the subsystems will account for 80 percent of the total cost (Pareto's Law of Distribution—Figure 6.1). The point is that project engineers should concentrate their time and effort on the high-cost subsystems in order to enhance their chances for establishing an accurate cost estimate.

6. *Failure to assess and provide for risks:* By its nature a prototype device involves engineering and design effort that must be breadboarded and bench tested for verification. Such tests usually involve an expenditure of effort to redesign and refine.

The project enginer must recognize the necessity to plan for the additional effort and to use judgment as to the number of man-hours and

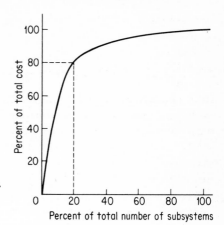

FIG. 6.1   Pareto's Law of Distribution.

additional material required for the risk areas in order to avoid under-estimating the cost of the equipment.

6.4   Work Breakdown Structure

The RLS will be used to demonstrate some of the cost-estimating principles and techniques previously presented.

The first step in establishing a cost estimate is to divide the project into segments consistent with the company's organization. In other words, the Acmen Electronics Company employs or has access to individuals and groups who are competent to engineer each of the technical areas shown in tier III of Figure 3.1. Further, the company has the specific resources to fabricate and test each of the subsystems noted above.

In making a cost estimate for designing the Shadow Generator Subsystem, the project engineer may assign the task of analyzing the scope of effort to the cognizant engineering group leader. This engineering effort would then be divided into two definable categories: development effort and design effort.

The characteristic of development effort is that it requires creative engineering wherein new areas are probed in an attempt to solve a particular problem. The development inherently contains more risk to complete as far as time and cost are concerned, and the estimates must therefore be modified by the proper risk factor.

The design effort involves straightforward engineering work in which established procedures are used to achieve the design objective.

The estimate of cost and time to complete the engineering work in the standard design area can be readily derived from the past experience of the company or from the history of previous jobs. The estimates should be accurate within 10 percent; therefore a small or zero risk factor would be adequate.

6.5 Classification of Engineering Effort

In a classification of the type of engineering effort that is required, each subsystem would be divided into discrete elements and analyzed. Figure 6.2 shows a block diagram of the basic elements which comprise the proposed Shadow Computer design. The elements identified in each block of Figure 6.2 serve as a logical basis for the work breakdown structure which would be used for deriving the cost estimate.

The elements of the Shadow Computer derived from Figure 6.2 are tabulated in Table 6.1. It should be noted that the Shadow Computer is identified as a tier III subsystem in Figure 3.1 (Shadows). Each of the elements is classified as to the degree of development required by the Acmen Electronics Company. It should be noted that the classification is based on the degree of development required by the particular company and that it will vary in accordance with the knowledge possessed by any company in a particular field.

In industry, a design or development that has been financed out of the company's funds is considered to be proprietary to that company. By contrast, it should be noted that when a development is financed by an outside party, such as the government, such a development becomes the property of the party who financed the development. In general, any development financed by the government will be made available to any other party having the proper security clearance and "need to know." Thus, in the case of the Shadow Start element of the Shadow Computer, if the element had been completed under some government program, it

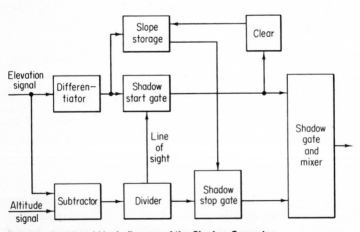

FIG. 6.2 Functional block diagram of the Shadow Computer.

TABLE 6.1 Detailed Direct Charges: Shadow Computer

Element	Class of effort*	Engineering hours	Breadboard and testing hours	Drafting hours	Material dollars	Manufacturing and assembly hours
Differentiator	B	40	· · ·	20	300	15
Subtractor	C	10	· · ·	15	400	10
Slope storage	A	160	· · ·	160	250	175
Shadow Start	A	150	· · ·	125	300	150
Divider	C	20	· · ·	15	500	15
Shadow Stop	A	200	· · ·	250	200	200
Clearer	C	10	· · ·	25	550	20
Gate and mixer	B	40	· · ·	10	300	15
System	B	50	300	50	300	100
Total		680	300	670	3,000	700

* A Development
  B Standard design
  C Off-the-shelf

would be available to Acmen Electronics for use on the project under discussion. If such were the case, the class of effort for the Shadow Start element would be C instead of A and the electrical engineering hours might be about 30 instead of 150.

It is therefore incumbent upon project engineers to explore all sources of information to procure available design data and thereby minimize the amount of costly development effort that would otherwise inflate their estimated proposal cost of any procurement for which they are bidding. (The United States government maintains an agency called the Defense Documentation Center which serves as a vast fountain of information for parties interested in any subject and possessing the required security clearance and need to know.)

Each of the Shadow Computer elements has been classified as to type of effort. As noted in Table 6.1, the slope storage, Shadow Start, and Shadow Stop elements classified as A require engineering development effort since there is nothing comparable to these elements which might be used by Acmen Electronics. Therefore the major portion of engineering effort would be required for these new functional elements.

The subtractor, divider, and clearer elements classified as C are commercially available or have already been designed so that they can be incorporated in the Shadow Computer system as they exist. These units require a minimal amount of engineering effort as noted in Table 6.1. The elements classified as B require some amount of engineering effort, primarily in the area of design modification to adapt their application to the overall system.

The hours required for drafting and manufacturing depends upon how much of the engineering and manufacturing effort will be done by the company. For those class C elements that will be used as off-the-shelf components, the only drafting that will be required is that which conveys how the particular element will interface with the complete system. The class A elements require a significant amount of drafting effort since such subsystems will be designed, built, and tested by the company. Therefore, detailed drawings will be required to enable other departments of Acmen Electronics to perform their functions based upon what is known in the drawings.

The estimate of material costs reflects a pattern that is opposite to the required hours for engineering, drafting, and manufacturing. The cost of an off-the-shelf element by its nature includes all the engineering, manufacturing, and other costs so that the estimated figure would be much higher than the costs for components that still have to be assembled, tested, etc. It is common practice for companies to treat all items purchased or obtained from different sources as material costs—even though some items may be fully assembled as operable components. Thus the class C items identified in Table 6.1 show an estimated cost significantly higher than the components that will have to be assembled to make up a class A element.

The analysis of the information in Table 6.1 supports the notion conveyed by Pareto's Law of Distribution, Figure 6.1, as far as engineering, drafting, and manufacturing man-hours are concerned. The three class A elements represent 33 percent of the items but account for about 77 percent of the hours required for their creation. The project engineer would be prudent to concentrate on the cost estimate of the high-cost class A elements to gain a better chance for deriving an accurate cost estimate for the Shadow Computer.

The engineering hours, drafting time, material costs, etc., as listed, reflect areas in which various lead engineers in charge of the tier III design effort indicated in Figure 6.2 are knowledgeable. In many companies, all drafting is centralized as a separate department, so that the estimate of drafting time would not be compiled in the same manner as the engineering effort. However, in such an organization, the drafting estimate is based on the information formulated by the design engineering department, and therefore the close correlation is maintained.

6.6   Management Review of Engineering Effort

The design and other costs are gathered and reviewed by progressively higher levels of management of the company. The functions of the various higher levels of management are to modify the estimates by eliminating duplications, to promote standardization, and to avoid excessive estimates of tare in material or hours. The project engineer is the key figure who interfaces with higher management officials during the review cycle.

When the estimates of engineering hours for various subsystems or elements are submitted by the different engineers, the review by the group leaders, project engineer, and higher management officials would include the following:

1. Review all systems to identify identical elements for which redundant engineering charges are estimated.

2. Review all systems to identify elements for which a design may have been accomplished on other projects, thereby making available an off-the-shelf design instead of expending a duplicating engineering effort on the current project.

3. Review all systems to identify elements which, although different, may be sufficiently similar to warrant adopting one standard element for a maximum number of systems without seriously compromising the performance characteristics of the hardware.

4. Review all systems to identify elements which may be sufficiently similar to an off-the-shelf design to warrant adoption of such an off-the-shelf design without compromising the performance characteristics in any significant way.

5. Review available sources of information to determine whether any design data relating to the project on hand is available for use by the company.

6.7 Compiling and Review of Material Costs

The estimate of material costs is a fairly straightforward effort, normally based on the current level of market prices. For projects spanning a long period, the material costs may be modified to anticipate the price that the company expects to pay at some future date. The modification factor is usually based on price indices and trends. However, for normal short-term projects (1 year or less) the current market prices are used. These costs must include factors to take care of waste, spoilage, and general fabrication errors normally referred to as tares. Thus, lead engineers would in their estimate of material quantities add a quantity to take care of tare, based on a statistical curve derived from past experience with a particular component.

The curve reflecting the spoilage of any particular element assumes the general shape of an exponential curve illustrated in Figure 6.3. The curves will assume different shapes for different elements and different histories of experience. For instance, curve a in Figure 6.3 might be for a

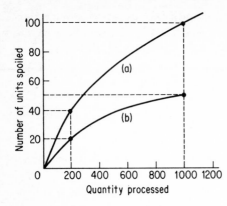

FIG. 6.3   Curves of spoilage vs. quantity processed.

type of transistor not generally used by Acmen Electronics. Therefore the spoilage rate on the basis of 200 units would be 20 percent. For 1,000 units, the spoilage rate would be 100 units or 10-percent spoilage.

The spoilage curve for a standard electronics component tube used more or less on a regular basis by Acmen Electronics might follow curve *b* of Figure 6.3. The rate on the basis of using 200 units would be 10-percent spoilage, and for 1,000 units, the rate would be about 5-percent.

The slope decrease or flattening out of any curve as the quantity of units used is increased is due in part to the experience or learning factor that is realized, thereby cutting down the spoilage rate for larger quantities. This is a reflection of the philosophy that, if someone makes a mistake resulting in spoiling an element due to carelessness or other shortcomings, that person will make special effort not to repeat the mistake. Theoretically speaking, if a person performs a certain task with certain material often enough, a point will be reached where no mistakes are made or where the spoilage curve is horizontal.

Further, the shop supervisor and production manager will use the mistake experienced by one worker which caused spoilage to alert other workers handling the same element against duplicating the unfortunate experience. Other steps such as providing special jigs, fixtures, tools, and procedures would be instituted to provide for a more efficient and less wasteful operation.

6.8   Estimating Manufacturing Costs

When the estimated engineering effort and material costs are gathered and correlated by project engineers, their next step is to determine the estimated manufacturing costs for the project. The production manager would be the key individual for estimating manufacturing costs, and the detailed procedures for gathering such costs would vary for different organizations.

Essentially, the procedures for getting estimated manufacturing costs would involve the following basic functions:

1. Providing the production manager with information relating to the hardware to be produced
2. The establishment by the production manager of the basic areas of effort
3. The solicitation by the production manager of cost estimates of different production supervisors
4. Estimation of effort by the different production supervisors
5. Accumulation and modification of production costs by the production manager
6. The transmission of cost estimates to the project engineer

The type of information required by the production manager would be a comprehensive description of the physical design of the different systems, their relation to each other and the overall hardware. In addition, a description of the specification requirements which relate to the hardware must be provided and interpreted for the use of the production manager. This is particularly important if the equipment must be subjected to environmental tests such as vibration shock since special manufacturing processes would be involved.

In the case of the Shadow Computer for the RLS, preliminary sketches of the various elements such as the differentiator and shadow start gate would be provided and described. In addition, a list of components such as amplifiers would be given.

The production manager would categorize all the types of production effort and disseminate the information to the various shop supervisors for detailed production estimates. Examples of categories of manufacturing effort would be assembly wiring, sheet metal work, and cabling.

The shop supervisor of each production area will, based on the descriptions given by the production manager, estimate the production hours required for this area of responsibility and transmit these figures back to the production manager who, in turn, will correlate all the figures and derive an estimate for the number of production hours required to complete the project.

The cycle by which the production hours are derived is completed when the production manager provides the project engineer with these figures.

6.9 Application of Overhead Rates

The different overhead rates and G&A (general and administrative) rates are applied to the direct costs that have been gathered and modified to evolve an estimate for the total project cost.

In Table 4.2 is shown a typical organization of the major cost elements of a bid proposal of two of the contracted items. The total cost would assume the same form but will summarize the individual items. The overhead and G&A rates are a function of the past experience of the company and will vary significantly among companies, geographic areas, and industries. The rates are adjusted periodically to reflect the latest company experience.

It should be noted that a company operating at full capacity will experience minimum overhead rates, whereas a company which has a significant amount of excess capacity will experience high overhead rates. The reason for this is that overhead burden is essentially fixed and that the costs such as heating, lighting, and real estate taxes are prorated over the in-house business. There have been cases of companies having essentially one job in-house that consumes a fraction of the available capacity so that the entire overhead burden was loaded on the one job. In such cases, the overhead rates are extremely high. Basically, the subject of the establishment of rates for overhead, G&A, etc., is a matter for accountants and auditors. Project engineers need only to be aware of their existence and their effect on the project. However, there is very little that they can do to influence them.

6.10   Commercial Considerations

Theoretically speaking, the estimated project cost (including profit) would be identical to the target bid price for a project. However, such an ideal situation rarely occurs since in the free competitive market, the commercial considerations must be recognized even when bidding on a government procurement open only to a limited number of qualified companies.

The final revision of the derived project costs is dictated by such commercial considerations and is made by the highest management echelons of a company. These considerations will involve too many facets to discuss separately in detail. However, the general categories of consideration that come into plan are covered in the following breakdown: type of effort required, type of contract solicited, type of competition, and company's need for additional business.

The degree to which the management is willing to modify the estimated project cost downward is a function of the risk of sustaining a loss on the contract that the company is willing to assume by submitting a low bid. On the other hand, the management of a company may, for different reasons, feel that it enjoys a position of advantage in a procurement and therefore will be willing to assume the risk of not getting a contract award by raising the estimated project cost in order to realize a greater profit. This basic philosophy, however, is contaminated by crosscurrents of other factors such as incentive and penalty clauses and profit sharing. The basic principle should, however, be kept in mind.

The risk factor associated with different types of contracts is discussed in Chapter 7. To cite the two extremes, a company on a cost-plus-fixed-fee procurement can afford to reduce its cost figures by a significant amount without assuming any risk or loss since all the incurred costs are reimbursable. On the other hand, if the procurement is a straight fixed-price contract, the maximum risk would be incurred if a reduction in cost is made.

The type of competition will help determine whether the estimated cost figure should be revised. If the Acmen Electronics Company has reason to believe that a competitor with a history of submitting low prices on procurements is one of the companies solicited and if the procurement is considered to be worth the risk, its management may shade their costs to a point which is estimated will be competitive with the anticipated price of the order company. The company's existing plant loading, including engineering work, will be a determinant as to how much risk the company would be willing to assume in order to obtain additional business. Basically, a variation of the forces of supply and demand enters the picture. If there is excess capacity in the plant, a company should be willing to assume more risk by adjusting its bid downward than if all the facilities of the company were being operated at full capacity.

In an analysis of how much of an adjustment is to be made in the estimated cost figures to arrive at a bid price, consideration must not only be given to the present commercial status of the company but to the status as projected for the future. For instance, the landmass simulator program is scheduled as a 65-week effort, and the work load demands on the engineering and manufacturing resources of Acmen Electronics that would have to be handled by the RLS contract are illustrated in Figures 4.2 and 4.3, respectively. The fabrication shop of the company may be fully loaded for the next eight months, but after that the current projection shows no significant work. It would be essential to the company to obtain the contract award for the landmass simulator in order to place work in the shop 8 months from the date of contract award as shown in the schedule indicated in Figure 4.3.

The same factors just discussed would be considered in establishing the ceiling and incentive arrangement to be included in a bid proposal. The actual mechanics of the fixed-price incentive contract (FPI) are discussed in Chapter 7. Any adjustments in the cost figures instituted by management must be reflected as adjustments in hours of effort or material costs in order to provide a consistent picture. It should be recognized that the adjustments by management of cost estimates that are deemed necessary to derive a bid price are very legitimate and necessary as far as the existence of the company is concerned. A downward adjustment in estimated cost would not be executed in a completely arbitrary manner but would be derived after consultation with the people in charge of the affected areas of effort. Usually, it is found that a limitation imposed on one's area of effort stimulates one's imagination and resourcefulness to seek cost-cutting steps in order to operate within set budget figures.

6.11 Summary

The very important task of making accurate cost estimates to serve as a basis for a proposal bid price is performed under the direction of the project engineer. The procedure for arriving at a cost estimate requires that the effort necessary to complete a project be divided into logical segments and that such segments be assigned to different engineering branches and production departments of the company's organization. Generally, the division of a project will have already been accomplished in conjunction with the completion of earlier phases of a project.

A previously designed and manufactured product that requires little or no engineering effort would involve the use of statistical data derived from past production experiences for establishing the cost estimate. A project, such as the RLS, requires the application of engineering and development effort; therefore the cost estimate would be based to a large degree upon judgment and experience.

The various categories of engineering hours, manufacturing hours, and material costs and estimates of what is required to complete the effort in different areas of the project would be handled by different group leaders.

The project engineer then accumulates all the figures and correlates them to eliminate areas of duplication and excessive estimates and takes advantage of every possible cost-saving factor.

Top management makes the final review of the estimated costs and adjustments to take into consideration the commercial, corporate, and competitive aspects of a procurement.

PROBLEMS

The Science Manufacturing Company is working on a procurement to develop and design a new ignition system for gas clothes dryers. The disciplines involve the design, manufacture, and assembly of a system comprising a heat-sensing element, a solenoid-operated gas valve, an igniter, and the timing circuits. The block diagram evolved for its engineering analysis is as follows:

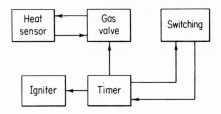

The following are pertinent facts relating to the Science Manufacturing Company which would be related to its cost estimate:

1. Experience in design and manufacture of timing circuits and switching and solenoid valves

2. Little experience in heat-sensing and igniter systems

3. Anticipated contract award date: January 1, 1965

4. Engineering department loading: 100 percent until December 1, 1964; decrease at rate of 25 percent per month after December 1

5. Manufacturing plant loading: 90 percent until February 1, 1965; decrease at rate of 20 percent per month thereafter

6. Type of contract solicited: FPI

7. Competition: very high

8. Rates and Burdens:

Engineering	$4 per hour	Burden 125%
Manufacturing	$3 per hour	Burden 175%
G&A Rate	10%	
Profit	10%	

Analyze the procurement requirements and develop a cost estimate taking into consideration all factors cited.

Explain the logic of the steps leading to the final proposal bid price.

Chapter Seven

Types and Analysis of Contracts

7.1 Basic Contract Concepts

There are two basic types of contracts used in the procurement of hardware: fixed-price and cost-reimbursable.

In the fixed-price contract, the cost is established at its signing and remains fixed provided that no changes in requirements are made and the contractual commitments made by the customer are upheld.

In the cost-reimbursable contract, the contractor is reimbursed for all or part of the costs incurred in the execution of the contract.

Within the framework of each of the two categories, many variations of contracts exist, each of which is applicable to the circumstances that are related to a particular procurement. The major types of variations will be described. The reader is again reminded that whereas these types of contract are more commonly used for government procurement, they also can be, and often are, used for procurements by corporations and other commercial firms. When reference is made to customer, the application could be to either the government or a commercial firm.

7.2 Determining Factors for Types of Contracts

Many factors influence the type of contract to be used for a particular procurement. Some of these factors are:

1. Complexity of design and type of equipment that is required, e.g., a highly complex development design would require a cost-reimbursable type of contract.

2. The amount of risk which the contractor would assume, e.g., a procurement requiring research effort in new areas involves considerable risk and would normally call for a cost-reimbursable type of contract.

3. The period of contract life, e.g., a procurement that extends over a long period would be subject to variables such as rate changes, overhead variations, and other unpredictables. A redeterminable type of fixed-price contract which would protect the contractor against increased costs and provide the customer with any credits when decreases occur would be normally used.

4. Competition present for a solicitation: if a procurement which might normally be the subject of a cost-plus-fixed-fee (CPFF) type contract interests a number of companies, the customer can solicit a fixed-price type of contract, which is more desirable from the customer's point of view.

5. Other factors such as demonstrated performance of contractor, difficulty in estimating costs of a procurement, urgency of requirement, and contractor's accounting procedures would enter into consideration and would influence the type of contract to be adopted.

7.3 Fixed-price Contracts

The fixed-price contract reflects an agreement by which the contractor is obligated to deliver the items as described in the specification and the contract schedule for a specified price. Precluding any changes in the contract or specification requirements, the contract price will remain fixed for the life of the contract.

However, for any major procurement involving engineering and/or development there are almost always factors which act to mar the theoretically perfect fixed-price contract and require consideration as far as effect on contract price is concerned. Some of these factors will be discussed in more detail.

The basic fixed-price contract is referred to as a firm fixed-price contract. Whereas a firm fixed-price arrangement imposes the greatest risk, it also offers the contractor the incentive and opportunity to realize the greatest profit. In effect, the contractor shares 100 percent in any savings due to cost reductions resulting from his efforts. The firm fixed-price contract is applicable when the purchased item can be identified in detail and when price competition exists. If any technical unknowns exist, the firm fixed-price contract is not an attractive vehicle for a contractor.

The variations of the fixed-price type of contract are fixed-price with escalation, fixed-price with redetermination feature, and fixed-price incentive.

Fixed-price with Escalation Contract: The fixed-price with escalation type of contract is used to protect the contractor against increases in labor rates, material, overhead, and other costs that are involved in performing under the contract. The escalation feature is usually tied to some official index such as labor rates for the area, basic raw material prices, and other indices. This type of contract is applicable when very long delivery schedules are involved.

Fixed-price with Redetermination Contract: The fixed-price contract with the redetermination feature is applied where large quantities of complex equipment are procured. Its use is justified in cases not only when the engineering and design effort is difficult to estimate but also when the production effort and materials for the large quantity of deliverable items are indeterminable. The terms of the contract can be worded to protect the buyer or the seller or both. Probably the most widely known examples of the redeterminable fixed-price contracts are those instituted during World War II when the armed forces required that complex weapons be designed and produced in large quantities on an urgent delivery schedule. In many of those cases, the United States government conducted redetermination negotiations for years after the conclusion of hostilities and the termination of the contracts.

There are several variations of the fixed-price contract with redetermination clauses. The following summarizes some of the different conditions that can be incorporated in this type of contract:

1. *Upward and downward adjustment:* Redetermination proceedings are conducted to establish a more accurate contract cost figure based upon cost experienced after the contract has run for a significant period of time. Both the contractor and the customer are afforded protection with this arrangement.

2. *Downward adjustment only:* This type of redetermination is known as a maximum-price contract and protects the customer. The application of this arrangement occurs when the contractor has been permitted to incorporate large contingency figures in the contract price. Any redetermination negotiations that are conducted seek to identify which of the contingencies were valid.

3. *Redetermination during the life of the contract:* As a general rule, it is

desirable to establish an accurate contract price as soon as feasible. The time of redetermination proceedings can be established as of a specific date, a particular point in the contract life, or any other agreed-upon arrangement. In essence, what is desired is to establish a time in the contract life when sufficient cost figures have been accrued which can be used for establishing a contract price. An example of 2 and 3 above might be where a contractor had included a contingency to provide a 50-percent redesign effort. At the conclusion of the design, if it were found that only a 25-percent redesign was necessary, the redetermination would reduce the contract cost by an amount equal to the 25-percent unconsumed engineering redesign cost. Further, because only half of the redesign contingency was utilized, it would be logically deducted that less drafting or laboratory development time would be required and that an anticipated reduced effort and cost in these areas could be eliminated, thereby reducing the contract.

4. *Redetermination after completion of the contract:* Since redetermination negotiations are usually involved and time-consuming, the customer or contractor may not be able to enter such negotiation during the life of the contract, e.g., in time of war. Therefore, redeterminations are stipulated for some postcontract period.

Other variations of the fixed-price with redetermination features can be implemented to fit the requirements of the procurement and the times.

Fixed-price Incentive Contract: The fixed-price incentive (FPI) is a form of contract used for major procurements which extend over a long period of time and/or when large production quantities are involved. Its aim is to reward or penalize contractors based on their ability to control costs primarily in manufacturing and general administration. The FPI contract is not applicable where a significant amount of research and development (R&D) effort is required or where the major cost element is for materials.

The FPI contract is set up in such a way that a target cost and target profit (generally 10 percent) are established. In addition, a contract ceiling price and price adjustment formula are established during negotiations between the contractor and the customer.

The understanding of the somewhat complex FPI contract can be enhanced by setting up a hypothetical case such as the following:

Target cost $100,000
Target profit (10%)   10,000
Ceiling cost (125%)   125,000

The cost-sharing arrangement is 80–20 above and 90–10 below the target cost. Assume that the contractor experienced an overrun of 20 percent in the contract cost. Under the terms of 80–20 for overrun costs cited above, reimbursement would be broken down as follows:

```
Target cost .........................  $100,000
Plus 80% of 20,000 (Portion of
   overrun assumed by customer)  ....     16,000
   Total reimbursement costs ...................  $116,000
Target profit .......................  $ 10,000
Less 20% of 20,000 (Portion of
   overrun by contractor) ...........      4,000
   Total profit................................        6,000
Total reimbursement .........................      $122,000
```

$$\text{Percent profit} \ldots\ldots\ldots\ldots \quad \frac{6,000}{116,000} = \text{about } 5\%$$

If, on the other hand, the contractor completed the contract, incurring costs of $80,000 ($20,000 less than target cost) under the terms of the cost-sharing arrangement of 90–10 below target cost, the following would evolve:

```
Target cost .........................  $100,000
Less 90% of 20,000 (Customer's
   portion of saving) ................     18,000
   Total reimbursed costs ......................   $82,000
Target profit .......................  $ 10,000
Plus 10% of 20,000 (Contractor's
   portion of saving) ................      2,000
   Total profit................................       12,000
Total reimbursement .........................       $94,000
```

$$\text{Percent profit} \ldots\ldots\ldots\ldots \quad \frac{12,000}{82,000} = \text{about } 15\%$$

In no case will the contractor be reimbursed in an amount exceeding the 125-percent ceiling ($10,000 × 125% or $125,000). The $125,000 ceiling figure includes all costs incurred, including profit.

It should be noted that the realized profit is expressed as a percentage of the cost of a contract.

7.4   Two-step Procurement

The two-step procurement is characterized by the following features:

1. The first solicitation is made solely for technical proposals based on a performance specification.

2. Technical discussions are held with those offerors whose pro-

posed design approaches are feasible and responsive to the specification requirements. The primary intent of the technical discussions is to bring all proposals that are under consideration to the same level of competence as far as meeting the performance requirements of the specification.

3. The second step of the procurement is the solicitation for bids. Award of a fixed-price contract is made to the offeror submitting the lowest bid.

4. The offerors' technical proposal upon which the successful bid is cited is one of the contractual documents and in some cases can take precedence over the performance specification.

5. At least two offerors participate in the procurement.

The two-step procurement is used when the following two circumstances exist:

1. There is insufficient technical data to prepare a design specification so that offerors must prepare and submit technical proposals. (This circumstance is also the basis for a negotiated procurement which generally leads to an incentive or cost-reimbursable contract.)

2. A fixed-price contract is desired.

The two-step procurement offers the best of advertised and negotiated procurement worlds—namely, a fixed-price contract in a procurement environment that normally calls for an incentive or cost-reimbursable contract.

7.5   Cost-type Contracts

The cost-type contract imposes the greatest risk on the customer and is used primarily in procurements involving significant R&D effort. Because of the risk imposed on the customer, it is essential that the contractor selected be established and have a record of proven integrity and reputation. Inherent in the cost-type contract is the intent and execution of the best efforts by the contractor in completing the project. If the all-important element of contractor integrity is deficient or lacking, then the customer will suffer since the cost-type contract does not offer any monetary incentive for efficient operation.

There are many versions of the cost-type contract but the two most widely used are the cost-plus-fixed-fee (CPFF) and the cost-plus-incentive-fee (CPIF). These are the cost-type contracts used for procurements from a commercial contractor as contrasted with procurements from a nonprofit organization such as a university.

The CPFF contract establishes a fixed fee or profit based on some percentage of an anticipated contract cost negotiated by the customer and contractor and stipulated in the contract. The fee usually is about 7 percent of the stipulated contract cost. The contractor is reimbursed for all allowable costs that are incurred in completing the contract. Thus, if the contract cost is $100,000 with a fee of $7,000 and the contractor incurs costs of $200,000, the fee should remain constant. There are situations when the contractor is entitled to fee on costs which exceed the contract cost. The more prevalent situations, when this occurs, are as follows:

1. Change in scope of contract, e.g., quantity of items and specification changes

2. Costs incurred which are beyond the control of the contractor, e.g., strikes, insurrection, and forces of nature

3. Adverse effects on contract due to the failure of customers to meet their contractual commitments, e.g., delivery of data and/or parts as required, late approval, or comments on design reports

The point that should be made is that contractors can obtain reimbursement for fees if excessive costs were incurred due to circumstances beyond their company's control.

Claims for fee on CPFF contracts go far beyond the amount of fee involved. The performance of a company is judged on whether the acceptable products are delivered on schedule and without cost overrun. Therefore, most contractors are anxious to maintain a good reputation by minimizing or eliminating the stigma of experiencing cost overruns. A cost overrun represents an amount by which the accumulated costs exceed the stipulated contract cost of the contract. Because such costs are unexcused (due completely to the fault of the contractor), they indicate how well the contractor's project engineer has performed in managing the contract. A chronic history of cost overruns by any contractor is a very good indication that the contractor has a poor manager or is, in effect, in the wrong business.

The CPFF contract is used for procurements which involve R&D or for delivery of a first production item. The justification for its use is the fact that the costs (engineering, manufacturing, and materials) cannot be estimated with any degree of accuracy at the time of award. In many cases, when a performance specification forms the basis of the procurement, a CPFF contract is used. Because of the risk that the customer assumes, the CPFF contract usually incorporates different types of clauses which serve to enable the customer to exert close control over the performance of the contract. Some of the clauses include the following provisions: customer approves major subcontracts, customer obtains title of property prior to delivery, allowable costs are defined, and overhead rates are negotiated.

The CPIF contract combines some of the features of the incentive and cost-type contracts discussed previously. This type of contract provides for the reimbursement of all allowable costs incurred by the contractor plus a fee based on an incentive formula for performance. It is generally used as a compromise when the offeror is unwilling to enter into a fixed-price contract and when the customer does not want to have a CPFF contract.

One of the various types of contracts described above will be selected by the customer as the most feasible for a contemplated procurement. It must be recognized that the type of contract the customer would desire (e.g., a fixed-price contract) is not always feasible since it could be very possible that none of the qualified prospective contractors would be interested or willing to enter into an agreement which although desirable for the customer is too risky for the contractor. Therefore, a customer must analyze the contemplated procurement from many points of view and make a determination as to which type of contract is most favorable and at the same time would be acceptable to qualified companies interested in the type of procurement contemplated. When customers have some serious doubt as to whether their choice of contract types would stimulate an acceptable response from interested companies, an alternative type of contract is solicited. The primary objective is to select a type of contract that will promote competition for the procurement.

7.6 Analysis of Risk

The primary purpose for the existence of any commercial organization is to make a profit or as a minimum, not to suffer a loss. In certain special and unique circumstances, a company may knowingly enter into a contract which will result in a loss to the company, but such action is generally taken on the premise that the long-term results will result in a profit. Some examples of these unique circumstances are as follows:

1. The company will purposely bid a price that is lower than its projected costs because of knowledge of expectation of additional units to be procured in the future. The company will be willing to sustain the loss on the first contract in return for the engineering and production experience that would place the company in a favored position on the future procurement.

2. The company is willing to enter into a contract which will result in a loss in order to keep the key people of the working force intact during a slack period. Since certain fixed expenses are incurred regardless of the amount of business that the company possesses, the amount of the loss is usually equivalent to the prorated amount of the fixed expenses.

However, in its normal course of doing business, a company will enter into a contract only if potential profit justifies the risk that the contract poses. Project engineers must address themselves to the following two basic questions as regards risk that their company will assume.

1. Can the company deliver the equipment within the time frame specified in the contract?

2. Can the company design, manufacture, and deliver the equipment specified without exceeding the projected costs upon which the contract price is based?

In practically any endeavor, time is money. In the case of a contract, if unforeseen complications or mistakes require the expenditure of more time than was planned for any phase, additional salary expenses, overheads, and other extra costs must be charged to the contract. If a cost-type contract is involved, the customer will provide for the additional expenditures. If a fixed-price type contract is used, the contractor would be required to absorb the additional costs. Therefore project engineers involved in proposing and/or negotiating a contract must recognize the risk areas that exist in the project and provide whatever contingencies they judge are necessary in order to protect themselves. At the same time, they must recognize that they are involved in a competitive procurement and that they may price themselves out of consideration if excessive contingency factors were included in their price quotation.

For the RLS procurement, an analysis of the contingency factors for a FPI contract is presented and summarized in Table 7.1.

TABLE 7.1   Percent Contingencies for a Fixed-price Radar Landmass Simulator Procurement

Area of effort*	Maximum contingency factor, %	Competitive contingency factor, %
Engineering hours:		
Data and planning	10	5
Preliminary design	20	10
Breadboarding	15	5
Detail design	10	5
Drafting	5	22
Test and checkout	20	10
Manufacturing:		
Planning	5	2
Components	7	3
Wiring and cabling	7	2
Assembly	10	5
Alignment	20	10
Materials	20	10

* For CPFF-type contracts contingency not required.

The rationale upon which the contingency factors are based are as follows:

Engineering

1. *Breadboarding:* The breadboard by its nature is a trial-and-error operation by which an operating model of a particular system design is verified. In the procurement discussed, the breadboarding would involve such systems as the FSS, optical, readout, and similar systems.

2. *Detail design:* Although the detail design is based on what has been proved by the breadboard, some redesign effort required as an outcome of the test and checkout results should be recognized.

3. *Drafting:* The drafting effort is directly related to the detail design effort and bears a proportionate contingency factor.

4. *Testing and checkout:* This effort involves somewhat the same type of effort as the breadboard except that revisions or changes involve changes in the detail design and drafting effort.

Manufacturing

The FPI contract requires that contingencies be provided to cover fabrication, wiring, assembly, and other costs resulting from design revisions and changes, excessive spoilage, and other similar costs.

Materials

The contingencies for materials cover costs for components, raw materials, and subcontracted assemblies that might be required over and above those normally anticipated because of redesign and other adverse factors.

The magnitude of the contingency factors would, to a large degree, be derived from the statistical records of costs incurred on previous contracts that were due to mistakes and errors and could be expected on the procurement at hand. For instance, the statistics on drafting might show that 5 percent of drafting hours was due to correcting errors uncovered by the checker, and therefore the addition of 5 percent to the number of drafting hours would be considered as a normal expenditure to be contained in the estimate.

If project engineers provide for the maximum contingency factor as noted in Table 6.1, they may price themselves out of consideration for the procurement. They therefore must use their knowledge of the type of effort involved, the past performance of their company team, and the competition they face to arrive at a competitive contingency figure such as that cited.

7.7 Summary

The two basic types of contracts are the cost-reimbursable and the fixed-price contract. Under each type are many variations, each of which is best suited for a particular type of procurement.

The cost-reimbursable type of contract is used for procurements in which the effort and costs required to complete the contract cannot be accurately estimated. Such types of contract include those requiring significant R&D effort. Because the contractor is reimbursed for all allowable costs incurred on the procurement, there is little risk associated with this type of contract.

The fixed-price contract is one in which the contractor must absorb all or part of the incurred costs which exceed the contract target price. By the same token, the contractor is "rewarded" for all or part of the amount if the incurred costs are less than the contract target price. Thus, the contractor therefore assumes a greater risk on the fixed-price type of contract. Because of the different risk factors inherent in each basic type of contract, the company bidding on a procurement must evaluate such risks and adjust the bid price accordingly.

Figure 7.1 summarizes the degree of risk inherent in different types of contracts.

Risk on Contractor—Fixed Price				Risk on Customer—Cost Type			
Firm fixed-price	Fixed-price with escalation	Fixed-price with redetermination	Fixed-price Incentive	Cost	Cost-plus fixed-fee	Cost-plus-Incentive fee	Time and material
Degree of Risk ⟵				Degree of Risk ⟶			
Where Applicable							
1. Design specification 2. Competition 3. No negotiations	1. Design specification 2. Escalating labor and material costs 3. Negotiated	1. Quantities 2. Long delivery 3. Fluctuating labor and material costs 4. Specification not definitive 5. Negotiated	1. Performance specification 2. Negotiated 3. Fluctuating labor and material costs 4. Risks	1. Performance specification 2. Research and development 3. Negotiated	1. Performance specification 2. Development 3. Negotiated	1. Performance specification 2. Development 3. Negotiated 4. Possible cost reduction	1. Performance specification 2. Emergency 3. General services

FIG. 7.1 Types of contracts based on risk.

PROBLEMS

1. Prepare a cost estimate for a FPI procurement which consists of the following cost elements. The proper contingency factor should be added to each cost element:

Engineering labor	1,500 hours
Engineering hourly rate	$5 per hour
Overhead rate	150%
Manufacturing labor	2,000 hours
Manufacturing hourly rate	$3 per hour
Overhead rate	200%
Material	$5,000
General and administrative rate	12%
Profit rate	10%

2. The contract, based on the figures of problem 1, was awarded with a 120-percent ceiling and a split of 75–25 above target cost and 80–20 below target cost. If the contractor incurred costs of $50,000, what would be the amount of reimbursement to the contractor including profit?

3. If the incurred costs amounted to $35,000 under the above contract, what would be the amount of reimbursement to the contractor including profit?

4. If the incurred costs amounted to $75,000 under the contract discussed above, what would be the amount of reimbursement to the contractor?

Chapter Eight

Contract Clauses

8.1 Definition

Any organization such as a commercial corporation or government agency has procurement policies which are expressed as standard or boiler plate contract clauses. These clauses usually exist as printed pages and are attached to the particular applicable contract.

Since the boiler plate clauses set forth the operating ground rules for a procurement and describe the rights and obligations of all parties to a contract, it is highly important that the project engineer know the details for all their terms and how they affect the procurement under consideration. A brief description of those clauses with which project engineers must be concerned and how they apply to the procurement of the RLS will be covered. The clauses to be discussed are typical of those used by the government for procurement contracts but are also applicable, to a large extent, in commercial-type procurements.

It should be noted that if any clause of the contract schedule and boiler plate are in conflict, the language cited in the contract schedule would take precedence.

8.2 Changes Clause

This clause is applicable to both the cost-reimbursable and fixed-price type of contract, and the wording is essentially the same. By terms of the changes clause, the customer may, without notice, make certain changes in the contract requirements if such changes are deemed by the customer as not constituting a change in scope. If the changes affect the cost, delivery or other areas of the contract, the contractor must serve written claim which serves to initiate a review, negotiation, and an equitable resolution of the claim. In situations where a change requiring additional funds is directed, it is presumed that the customer, be it the government or private organization, has the budgeted adequate money to pay for the change.

When procurements are under consideration, the project engineer must not lose sight of the possibility that customers can exercise their rights in the changes clause at any time and the contractor has a primary obligation to comply. Whereas it would be unreasonable to expect the project engineer to possess clairvoyant powers and predict changes that will be issued (although in some cases, intimate knowledge of the procurement and very close liaison by the project engineer and the marketing staff with procurement officials can provide information from which changes can be anticipated), the project engineer should direct a design which is as flexible as possible to permit changes, if directed.

Changes can be directed to effect what amounts to a decrease in scope. Decreases in scope as a result of the application of the changes clause are relatively rare but the same principles and procedures as those discussed apply in reverse.

The changes clause of a contract is intended to give the customer the flexibility in satisfying the equipment requirements and to provide for improved features or performance characteristics which were not envisioned at the time the contract specification was adopted. The clause is not to be viewed by the customer's project engineer as a convenient tool for rectifying errors that result from a poorly conceived and written specification.

Changes to a contract, particularly a government contract, can assume the form of a verbal or informal direction. Often a change to a contract may be unintentionally directed and can end up as being contractually binding. Changes of this type are identified as constructive changes and are discussed in Chapter 10, the Legal World of Project Engineers.

8.3 Allowable Costs

The boiler plate clauses for cost-reimbursable contracts provide for controls and approval by the customer of the costs that are accumulated. In general, only those costs incurred in connection with the contract will be allowed and reimbursed. The criteria for determining what constitutes an allowable cost for government contracts is defined in the procurement regulation documents. For contracts between private parties, the terms of the contract generally define what constitutes an allowable cost.

An allowable cost is generally defined as one which would be considered as reasonable and applicable for the contract in question. A cost which would be incurred by a prudent person as a result of sound business practices would be considered reasonable. If, for instance, the contractor purchased subassemblies from another company as a noncompetitive subcontract and the subcontract costs were judged to be unreasonably high, those costs might be judged as not allowable by the customer's auditor. Discussions between the contractor and customer would be required to establish mutually agreeable figures, if possible.

An allowable cost must be applicable to the contract being changed. Engineers or other employees must charge their time to the project on which they are involved. Therefore, the contractor must have a system which will provide an accurate means of charging costs to the appropriate project.

The allowable-cost clause also sets forth the controls on the payment of profit or fee and the schedules of payment of a cost-reimbursable contract. The contractor is obligated to advise the customer at a predetermined time of any indication that the cost stipulated in the contract will be exceeded. In no case will the contractor be reimbursed for costs exceeding those in the contract unless they are authorized by the customer as a contractual price change.

8.4   Inspection and Correction of Defects

For both fixed-price and cost-reimbursable types of contract, the contractor is obligated to correct any defects uncovered prior to acceptance. In addition, the contractor is usually required to correct any defects uncovered up to 6 months after acceptance of the equipment. The contractor is required to provide "reasonable" facilities and assistance to the customer when the equipment is being tested. The customer usually has authorization to perform whatever tests or inspections that are deemed practical prior to the completion of acceptance tests.

For the cost-reimbursable contract, the contractor will receive payment for direct costs incurred as specified under the allowable-cost clause, although payment of fee will not usually be allowed at that time. The exception to the reimbursement of costs in the case of uncovered defects is when fraud or similar practices on the part of the contractor is revealed.

For the fixed-price contract, the contractor must remedy defects uncovered up to 6 months after acceptance, with no reimbursement. If the contractor fails to complete the necessary corrections, an equitable sum can be withheld by the customer. Any disagreements would be handled in accordance with the terms of the disputes clause.

In preparing the cost proposal for a procurement, the project engineer must provide for the time, facilities, and costs that will be required for the inspection and the corrections that would be necessary under the proposed contract.

8.5 Subcontracts Clause

About the only restriction placed on subcontracting on a fixed-price contract is that wherever possible, subcontracts be awarded as a result of competitive bidding.

In the case of cost-reimbursable contracts, the subcontracts clause affords the customer some very definite controls regarding the subcontract awards and effort. By the terms of this clause, the contractor must advise the customer of contemplated subcontracts. Contingent upon the size and type of subcontract that is under consideration, approval of major subcontracts must be received from the customer. However, consent of the customer to use a particular subcontractor does not relieve the contractor of the responsibility for meeting the terms of the contract in the event the subcontractor failed to perform.

In the event of any contractual difficulty such as slippages, claims, or other disagreements, the subcontractor is obligated to report such difficulties to the prime contractor as soon as possible.

It has been indicated that the Acmen Electronics Company would probably subcontract the detector and other systems. It will be the responsibility of the project engineer to draft a set of specifications, solicit subcontract proposals, and participate in the negotiation of the best possible contract. It would be to the interests of Acmen to enter into a fixed-price subcontract with their vendors.

8.6 Termination Clause

The two basic types of contract terminations that can be applied to either fixed-price or cost-reimbursable contracts are: termination at the convenience of the customer or government and termination due to contractor default.

A convenience termination is one in which customers decide that the material under procurement is no longer required and that they are prepared to assume whatever losses are associated by termination. In cases of convenience terminations, the contractor ceases all work and issues cancellations on all purchase orders and subcontracts immediately upon receipt of the termination notice. Within a specified time (usually a year) the contractor must submit the termination claim. The termination claim includes all incurred costs including cancellation charges on orders and subcontracts, special expenses associated with the termination effort, and other similar costs.

For the cost-reimbursable contract, the costs that are claimed are verified as applicable, and the fee is negotiated. If an impasse regarding claimed costs or fee develops between the contractor and customer, the matter is referred to higher authority for resolution (legal courts for commercial customers, and the Settlement Review Board if the customer is the government).

For a fixed-price contract legitimate costs and fees will be determined and negotiated when a convenience termination is executed. If the evidence of a procurement indicates that the contractor would have exceeded the ceiling contract price (thereby experiencing a loss on the contract), a sharing of the loss by the customer and contractor is negotiated. Again, the customer has inherent rights of appeal as cited above.

Termination by default is applied in contracts because of the failure of the contractor to deliver the item or items under procurement as specified or when there is concrete evidence of the failure of the contractor to make progress in the program, thereby supporting the supposition that the terms of the contract will not be met. The process in such terminations involves officially notifying contractors of their lack of progress which, in turn, requires a response within a stipulated number of days. If the response is not made or is unsatisfactory, the termination action can be executed.

In cases of default terminations of cost-reimbursable contracts, all substantiated costs are reimbursed. However, fee will be paid only on those items that have been accepted. Thus, for an item of the contract which is 90 percent complete, only those costs associated with the item will be paid with no fee.

The procedures and reasons for default termination on fixed-price contracts are the same as those discussed for cost-reimbursable contracts. However, the contractor will be paid only for those items that have been accepted. In the case of the above noted item that was 90 percent complete, the contractor in a fixed-price contract will receive no reimbursement for the item. Thus, the contractor stands to experience a complete loss on any outstanding items that have not been delivered. In addition, the customer has the right to procure the unaccepted items from another source, and the defaulted contractor is then legally obligated for any costs which the customer experiences over and above the original cost of the items involved.

8.7 Excusable Delays

In cases of cost-reimbursable contracts, delays are excusable if caused by circumstances beyond the control and without the fault or negligence of the contractor. Such causes would include fires, floods, or other similar causes that are identified as "Acts of God." Strikes, insurrection, or other human-created problems would also be identified as excusable delays.

The same type of excusable delays apply to subcontracts on a project, providing the following additional conditions are met:

1. The contractor had diligently monitored the subcontract to establish that the difficulties existed or were developing.

2. The contractor had diligently attempted to seek alternate sources for the subcontracted items but was unsuccessful.

3. The subcontractor did not contribute to the cause of the delay either by action or inaction.

A delay may be considered excusable if caused by the failure of the customer to execute some required contractual obligation. In such a case, the contractor may be entitled to receive reimbursement for expenditures due to the delay.

8.8 Disputes Clause

The same disputes clause language and terms are applicable to both the cost-reimbursable and the fixed-price contract. Any dispute (other than one involving allowable costs) not resolved by agreement is arbitrarily decided by written notification by the customer. The clause concerns questions of fact. After the decision is rendered by the customer, the contractor, within a specified time, can appeal. In the case of government contracts, the appeal is made to the Secretary of Defense (Navy, etc.), within 30 days. For commercial procurements, the appeal can take the form of legal action in an appropriate court of law.

If the dispute involves issue of allowable cost on government contracts, the contractor can appeal to the government auditor within 60 days after the determination by the government. Adverse decisions can be further appealed to the cognizant cabinet secretary.

For commercial procurements, the contractor can appeal to courts of law.

For government procurements, the relief that a contractor can seek is well defined by the universally applied disputes clause. For commercial contracts, the contracting parties should agree to procedures for resolving disputes and incorporate the specific agreement in the contract. Pending final resolution of any dispute, contractors have the obligation to pursue in a diligent manner the execution of their contractual obligations. This is specifically stated in the clause for government contracts.

8.9   Customer (or Government) Furnished Property

The same conditions relative to Customer Furnished Property are applicable to cost-reimbursable or fixed-price contracts. The government is contractually obligated to deliver the specific items by the dates set forth in the contract or, in lieu of specific dates, at a time sufficiently early to permit contractors to utilize the property in time to fulfill their obligation. If the property is not furnished as required, the contractor must serve written notice of the delinquency so that an equitable adjustment in cost and delivery can be made in accordance with the provisions cited in the changes clause. In addition to the responsibility that the customer has for timely delivery, the property must be suitable for use in the contract. It should be added that the Customer Furnished Property can include data, reports, drawings, and other materials as well as equipment.

Title to the property remains with the customer; however, the contractor has the contractual responsibility to maintain, repair, and preserve the property in accordance with sound commercial practices. If major failure or damage to Customer Furnished Property occurs which is not due to any negligence on the part of the contractor, the customer is obligated to make the necessary repairs or replacement and is liable for the responsibility of any schedule slippages or incurred costs that may result.

Property furnished by the customer may be agreed upon to be in "as is" condition. In effect, in such an arrangement, the contractor agrees to utilize equipment for the contract that may not be in a suitable condition for its intended use. The contractor then assumes the responsibility for making the necessary repairs, etc., to place the equipment in a suitable condition. Generally, the contractor also assumes the responsibility for making any repairs or performing any maintenance that may be requested.

8.10   Patent and Copyright Infringement

The patent and copyright infringement clause is applicable to both the cost-reimbursable and the fixed-price type of contract. In the event that the contractor, in performing the contract, infringes on the rights assigned to a third party, and such party initiates infringement litigation, the contractor is obligated to immediately file notice with the customer of the litigation. The customer is liable for all infringement charges arising out of the performance of one of its contracts. The customer is also responsible for reimbursing the contractor for any evidence that is requested except in those cases in which the contractor has agreed to indemnify the customer against the claim being asserted.

When a contractor indemnifies the customer against infringement, the contractor is agreeing to assume litigation and possible penalties for any patent infringement that may occur during the execution of the contract. The general conditions associated with such indemnity include the following: the customer must notify the contractor of any infringement allegations, the contractor must be given adequate time to defend the allegation, and the infringement must not be the result of the customer's direction.

8.11 Filing of Patents

The clause relating to the filing of patents is applicable to both types of contract. The contractor may pursue a patent application for an invention developed under a customer or government contract. However, the contractor must grant to the customer an "irrevocable, nonexclusive and royalty free license" to use the patent as desired. In cases of government contracts, the contractor retains the right to use the patent or sell patent rights to other parties for use on commercial applications. Even if an invention has no commercial value, a company may still desire to expend the funds to procure a patent to serve as a prestige asset or to enable the company to dominate its particular field.

If contractors do not desire to obtain a patent on an invention developed on a government contract, they still are obliged to furnish the government with all the necessary information and disclosures so that the government may apply for a patent. In such a case, the patent would be assigned to the government. If contractors resist submitting disclosures to the government on developments deemed to be subject for patents, the government can withhold a specified sum of money in payment of the contract.

8.12 Overtime and Shift Premiums

Overtime and shift premiums are not allowable on cost-reimbursable type contracts unless specific written approval for such premium time is received by the contractor from the customer. Overtime and shift premium payments are extended if they are:

1. Required for emergencies due to accidents, equipment breakdowns, etc.

2. Required for indirect labor employees performing project management, administration, maintenance, etc.

3. Required to perform tests and functions that cannot be interrupted.

4. Required to lower overall costs to the customer.

For fixed-price contracts, the only restriction is that the customer reimburse employees working in excess of 8-hour days.

8.13   Value Engineering

The objective of a value engineering clause in a contract is to provide an incentive for the contractor to analyze the contract requirements for the equipment under procurement and solicit proposals for changes which will result in cost savings. The incentive is that if the value engineering proposal is accepted, the contractor will share in the cost savings resulting from the proposed change.

The basic conditions that are necessary to support a value engineering reward include the following:

1. The contract must include special provisions to permit the implementation of value engineering proposals.

2. The proposal must address a specific contract requirement that can be modified at a savings in cost without adversely affecting the quality and performance capability of the product under procurement.

3. The value engineering proposal must be accepted by the customer.

4. A specification and contract amendment to reflect the cost savings cited in the value engineering proposal must be executed.

Value engineering concepts are mostly applicable to procurements based in design specifications involving larger quantities of items. For instance, the Navy traditionally required fixtures such as signal lights used aboard ships to be fabricated out of bronze and machined to close tolerances. Some time ago, a contractor for the item proposed an aluminum alloy material that was to be cast to tolerances that were not as precise as specified.

The value engineering proposal included the following basic justifications for the specification deviations:

1. The close tolerances specified for the product were unnecessary for the product under procurement.

2. The substitution of the aluminum alloy for the bronze housing could be accomplished without, in any way, affecting the performance, life cycle, maintenance requirements, or any other requirements of the product.

After evaluating the value engineering proposal, the government accepted the proposal, revised the specification language, amended the contract, and shared with the contractor the savings realized by the changes that were implemented.

Value engineering contract clauses are generally not applicable to procurements based on performance specifications such as the RLS. In effect, the project engineer performed what amounted to a value engineering analysis in considering the alternative approaches for the technical proposal that have been described in Section 3.6 and Figure 3.4 of this text. If the dual-FSS system were specified in the contract and the contract provided for value engineering proposals, the Acmen Electronics Company, after they received the contract, could have made a value engineering proposal based on the less expensive single-FSS system, thereby sharing in the savings that represents the alternate design.

8.14   Other Common Clauses

In addition to the above major standard clauses, there are numerous other clauses in the boiler plate. Most of the other clauses cover areas which do not relate directly to the project engineer's functions and are therefore not discussed in this text. Some of these fringe clauses cover the following areas and are handled by different departments of the contractor:

1. Buy America Act
2. Convict labor
3. Payment for overtime and shift premiums
4. Nondiscrimination in employment
5. Insurance-liability of third parties

Table 8.1 summarizes the essence of the various standard contract or boiler plate clauses discussed and contains the main elements of each clause.

TABLE 8.1   Summary of Significant Boiler Plate Clauses for Cost-reimbursable and Fixed-price Contracts

Type of clause	Cost-reimbursable contract	Fixed-price contract
Changes clause	Unilateral changes within scope can be made without notice by customer. If change affects scope of contract, equitable adjustment in contract is made.	Same
Allowable costs	Incurred costs are subject to the approval of customer. Contract cost cannot be exceeded unless specifically authorized.	Not applicable
Inspection and correction of defects	Contractor provides facilities and corrective effort up to 6 months after acceptance. Costs in general are reimbursed.	Contractor provides facilities and corrective effort up to 6 months after acceptance. Costs not reimbursable.

TABLE 8.1   Summary of Significant Boiler Plate Clauses for Cost-reimbursable and Fixed-price Contracts (Continued)

Type of clause	Cost-reimbursable contract	Fixed-price contract
Subcontract clause	Approval by customer required for specific types and sizes of subcontracts.	Competitive bids required.
Terminations convenience	Notice served by customer. Contractor submits termination claims.	Same . . . If contract would have exceeded ceiling cost, loss is assumed by contractor.
Default	Issued when lack of progress is shown or when contractor fails in contractual efforts. Notice is issued to contractor and response required in specified time. Contractor reimbursed for costs incurred. Fee paid only for accepted items.	Same Contractor reimbursed only for accepted items with fee. Contractor not reimbursed for any costs involving unaccepted items.
Excusable delays	Delays beyond the control of the contractor are excusable.	Same . . . If delays caused by customer, contractor is reimbursed.
Disputes clause	If agreement not reached, customer renders unilateral decision which can be appealed. Disputes involving fact and allowable costs are handled and appealed differently. Contractor obliged to perform pending appeal.	Same
Government furnished property	Government obligated to deliver property as required in contract in suitable operating condition and is responsible for repair or replacement if major difficulties occur. Adverse effects on contracts due to failure to meet property requirements subject to reimbursement or rescheduling. Contractor responsible for normal maintenance and repair.	Same

TABLE 8.1 Summary of Significant Boiler Plate Clauses for Cost-reimbursable and Fixed-price Contracts (Continued)

Type of clause	Cost-reimbursable contract	Fixed-price contract
Patent infringement	Government liable for infringement of patents by contractor in performing under a contract. Contractor is reimbursed for information furnished by contractor for use in defending suit.	Same
Filing of patents	Contractor may file but assign all rights to government. Contractor can use in commercial application. If government files in lieu of contractor, contractor must furnish all disclosures.	Same
Overtime and premium shifts	Written permission required for utilization. Valid only in special emergency or circumstances that preclude standard 8-hour operation per day.	No restrictions provided employee is reimbursed.
Value engineering	Contractor may submit proposals if value engineering clause is included in contract.	Same

PROBLEMS

1. The XYZ Aircraft Corporation is under contract to design and build aircraft to be used for submarine warfare. Shortly after the date of contract award, the government invoked its rights under the changes clause and directed the following two changes:

a. Substitution of an improved model of a radar display which is electrically and mechanically interchangeable with the previous model. The radar display is being provided as Government Furnished Property.

b. Increase the gross weight requirements of the aircraft while maintaining the original performance requirements.

What plan of action should the XYZ Aircraft Corporation take as a result of the above directed changes?

2. The Ace Equipment Company experienced crippling damage due to a fire and as a result experienced serious delays in performing on a $5 million government contract. The government had accepted equipment costing $1 million. Of the remaining items to be delivered, the company has incurred costs of $2 million prior to the fire.

a. What would be the justification of a termination by default, and if such a termination were executed, what costs or losses would be incurred by the government and/or the company?

    b. What would be the justification of a termination of convenience, and if such a termination were executed, what costs or losses would be incurred by the government and/or the company?

    3. The CD Corporation has a contract to design and build a power-supply center consisting of several electric generating units and an intricate system of control, regulation, and switching equipment. The electric generating units are to be provided as Government Furnished Property.

The CD Corporation is experiencing major difficulty in designing the switching equipment and, as a result, is faced with a 6 month delivery delay in the 24-month contract delivery schedule and a monetary loss of $300,000 in their FBI contract.

Concurrently, the government was 3 months late in providing the generating sets to the contractor. The contract stipulates that the generating sets are not to be used in conjunction with the design or development phases of the contract.

The CD Corporation has submitted a claim for $300,000 and a 6-month extension in delivery based on the failure of the government to provide the generating sets as required.

As project engineer for the government, prepare an evaluation of the claim of the CD Corporation.

Chapter Nine

Negotiations

9.1 Objectives of Negotiation

The project engineer, having completed the technical proposal for the Landmass Simulator which was transmitted to the contracting officer of the procuring activity, must now prepare for the negotiations which can be anticipated after the proposals for all offerors are evaluated.

During the negotiations, the official representative of either party is often a contract administrator or a lawyer who specializes in contract law and negotiation techniques. The project engineers almost always participate since most of the discussions relate to technical areas, cost breakdowns, schedules, and related subjects which are the areas in which the project engineer is most knowledgeable.

In order to avoid making statements which may jeopardize the position of their company, project engineers must be sensitive to the implications of any information that they may convey during negotiations. They therefore must have an intimate grasp of the fundamentals of contract law and related subjects. It is particularly important that project engineers know the exact limitations of what their company will accept concerning the price, schedule, type of contract, and other parameters of the contract being sought.

Although the contract administrator is generally the official representative of the company, the outcome of the negotiations depends largely on the project engineer. In many instances, however, the project engineer may be designated the company's representative. Therefore, the text in this chapter will discuss the role of project engineers as company representatives.

For procurements involving large sums of money in which a limited number of proposals were solicited, all offerors are usually given "a day in court." When there are a large number of offerors, only those offerors who have submitted the lowest quotations in conjunction with acceptable technical proposals will be invited to discuss their offering and explore the possibility of reaching a contractual agreement. Any procuring party, whether it is the government or an industrial firm, will officially state that for procurements based on performance specifications, price is not the determining factor. Whereas this is basically correct, the low bidder who is not declared nonresponsive is automatically placed in a preferential position, and consideration must be given to this proposal, even though this technical approach may not be considered the best of those submitted.

The object of the technical negotiation is to explore and clarify questionable areas in order to establish the relative merit of technical approaches in different areas of the offerors and whether the technical approach of any offeror falls short of meeting the minimum requirements of the specification and is therefore unresponsive. In addition, the project engineer evaluating the proposals would want to clarify any possible misinterpretation of the specification which had resulted in a proposed technical approach which represents an unnecessary embellishment or one which may not satisfy the specification requirements that were intended.

The negotiation relating to the proposal bid price is to establish the basis for or justification of the cost elements in each area of the procurement, whether offerors fully understand what is required as evidenced by their estimates of cost in various areas, and whether there is consistency between the cost figures and the technical description in each area.

9.2 Definition of Negotiation

Negotiation is a term used very broadly and, in many cases, loosely. Negotiation, as used in modern procurements, is an art embodying sophisticated tactics and maneuvers by both parties. In the final analysis, negotiation is a procedure wherein the exchange of concepts is verbally accomplished in order to arrive at a meeting of the minds as to what is required and what is offered as far as the technical, schedule, and price elements are concerned. Another way in which a negotiation can be described is as follows:

> Procurement by negotiation is the art of arriving at a common under-standing through bargaining on the essentials of a contract such as delivery, specifications, prices and terms. Because of the inter-relation of these factors with many others, it is a difficult art and requires the exercise of judgment, tact and common sense. The effective negotiator must be a real shopper alive to the possibilities of bargaining with the seller. Only through an awareness of the relative bargaining strength can a negotiator know where to be firm or where he may take permissive concessions in prices or terms.

Not all procurements lend themselves to negotiations. On standard items in common use for which the costs are fairly well established, the buyer would award a contract to the lowest bidder. The project engineer, however, is involved in procurements which often represent the first of a kind and for which significant creative engineering effort is necessary. For such projects, the project engineer must be prepared to participate in rigorous procurement discussions.

9.3   Establishing Negotiation Parameters

The primary object of project engineers is to obtain the contract at terms which favor their company to the best possible degree. In planning for the negotiation, they must establish the points beyond which the contract would not be acceptable to them. There are many factors which enter the picture in deriving the point of acceptability of the contract. Some of these factors are graphically portrayed in Figure 9.1.

Figure 9.1 provides a graphic display showing the relative effects that changes to delivery schedules, technical requirements, or contract terms would have on cost to produce a product. The intersection of the accept-ability line with the price structure norm (point *a*) represents the price that Acmen Electronics Company proposes for the trainer as specified in the RFP. The acceptability line has been established based on the various considerations such as competition, desire of the bidder for the contract award, experience, and risk factors associated with the procurement. If during negotiations any changes to the procurement parameters are considered, the project engineer can refer to a chart similar to Figure 9.1 and derive an instant guide as to what impact a suggested change in any of the major procurement provisions may have on the cost structure and whether the change is practical.

The increments shown on the vertical scale would be expressed in dollars over and under the bid price that the contractor offered when the proposal was submitted. A figure that is less than the bid price would fall below the acceptability line and therefore be favorable to the user since it represents a cost reduction on the proposal. A price above the line would be favorable to the contractor.

The horizontal acceptability line would be scaled to reflect the following: time for use with delivery and price-delivery curves, degree of complexity for use with technical requirements and price-technical requirements curves, and degree of contract hardship for use with contract terms and price-contract terms curves.

To apply the curves of Figure 9.1 to the RLS procurement, point *a* would represent the position of the Acmen Electronics Company at the time the proposals are submitted. In the interest of clarification, assume the bid price at point *a* to be $500,000. Each horizontal increment above the acceptability line could represent an addition of $10,000 to the bid price, and each increment below the acceptability line represents a subtraction of $10,000.

In like manner, the delivery requirements can be scaled to weeks along the acceptability line. Point *a* represents the delivery requirements of 65 weeks as discussed in Chapter 3. Each horizontal increment could represent 2 weeks added to or subtracted from the 65-week schedule. To illustrate the application of the Figure 9.1, suppose the customer desired to accelerate the delivery schedule by 4 weeks and the consideration in this case is money. The project engineer, in referring to the curves of

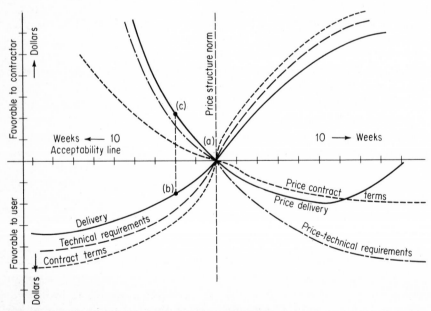

FIG. 9.1   Variations in price based on procurement revisions.

Figure 9.1 as a guide, can establish from the price-delivery curve that the contract price should be increased by $40,000 to permit the delivery accelerations. The figure of $40,000 is derived from the estimated cost of overtime, risks associated with accelerated effort, and other factors of this nature. The points in question are designated as b for the new delivery and c for the increased price. It should be noted that any point on the price-delivery and delivery curves must fall on the same vertical line in order that the correlation be valid and accurate.

Of particular interest is the slope of the two curves in question. The price-delivery curve slopes sharply upward when correlated with increasingly shorter delivery as expressed by the delivery curve. In other words, if the customer wanted the delivery improved by 8 weeks, there would be no practical point on the price-delivery curve that falls on the same vertical line as the 8-week improved delivery point, which means that the improved delivery is impossible as far as contract price is concerned.

The slope of the various curves are significant. For instance, in the favorable to user area, the delivery curve approaches a slope that is parallel to the acceptability line. The notion being conveyed is that beyond a certain point (approximately 12 weeks in this instance) any shortening of delivery time for the trainer does not result in any increased benefit to the user, even if such schedule shortening were possible or economically feasible. The same conclusion can be drawn by noting the slopes of the technical requirements and contract terms curve in this area.

It should be noted that the slope of the price-delivery curve reverses itself as the 65-week delivery schedule is extended beyond 8 weeks. Thus, up to a point, the contract price can be reduced if the delivery requirement is relaxed or extended. At the 65 plus 8-week point, however, Acmen Electronics anticipates that the landmass simulator will be ready for shipment, even as built at the leisurely schedule, and to keep the unit in the plant would result in the unnecessary accumulation of costs. It should be noted that the delivery curve slopes downward, indicating that beyond the 8-week extension, the delivery conditions cease to be in the favor of Acmen Electronics.

The curves identified as contract terms and price-contract terms, as well as the curves representing technical requirements, would be applied by the project engineer in a similar manner during negotiations.

The establishment of the price-delivery and delivery curves was relatively easy since these two curves could be derived directly from numbers such as weeks and dollars. However, in the case of the curve representing contract terms, the task is more difficult. How, for instance, could a more stringent requirement for reliability of operation be readily expressed in dollars except through a thorough design analysis? Since the curves are intended primarily as a guide to the project engineer during negotiations, the contract terms curve could express estimated degrees of hardship that would be imposed on Acmen Electronics, and the degrees of hardship could be correlated with the price-contract term curve. It is in this area where the technical knowledge, familiarity with the proposed contractual and technical details, and experience of the project engineer play a major role. If during negotiations the proposed contract terms are revised, the project engineer must be able to judge their impact on the effort of Acmen Electronics and thereby relate such impact in terms of price.

From Figure 9.1, the slopes of the contract term curve and the price-contract term curves indicate again that, at some point, no amount of money could compensate Acmen Electronics if the contract terms are made too severe. In like manner, there is a level at which the price-contract term curve becomes horizontal, indicating that no amount of relaxation of the contract terms would have an effect on the reduction of the procurement price.

The slopes of the technical requirements and the price-technical requirements curves are essentially the same as the contract term and price-contract term curve. Again, the technical requirements curve must be expressed in some convenient scale such as the degree of impact on the effort that is required in order to have the variations of the technical requirements expressed in the proposed contract price. As before, the project engineer must have the knowledge and capability to judge any proposed revision of the technical requirement in terms of price increase or decrease during negotiations.

The variations that could evolve during negotiations were discussed and presented in terms of the effects on the proposed contract price. In general, most negotiations would center around the question of how the price would be affected by changing some requirement. It is not uncommon for factors other than price to be the prime consideration. For instance, a family of curves which demonstrates how the delivery schedule might be varied can be drawn as different revisions in proposed contract requirements are suggested. If the technical requirements are tightened beyond a certain point, the delivery would be extended beyond a reasonable point and would be expressed by an upward sharp slope of the delivery-technical requirements curve. In other words, the technical requirements such as accuracy and response speed could be made so stringent that they could not be met, regardless of the amount of time that may be extended to the contractor to accomplish the task.

The curves of Figure 9.1 illustrate one convenient tool which can serve the project engineer in negotiating a contract. It should be emphasized, however, that in order to have the tools serve to the maximum advantage, their user must be thoroughly familiar with how they were derived and with the purpose for which they were designed. The reason for this is that for most of the curves the selection of a particular point is essentially the result of judgment. Therefore, if the project engineer using the curves during negotiations is the person who participated in their design, then the information and decisions resulting from their use will be the most accurate and the soundest. In addition, the very act of designing the curves forces the project engineer to study all the important technical contractual, cost, delivery, and other aspects of the procurement, thereby affording an excellent opportunity to review the procurement.

9.4 Analysis of Buyer's Position

A negotiation can be viewed as a contest in which parties will use to their advantage any situation that exists or that may develop during the proceedings. In any contest, one fundamental requirement is to study and learn everything possible about the opposition and their position.

In the case of the landmass simulator, the first obvious bit of information is that a specific requirement for the trainer exists. Further, the project engineer should note that the radar to be simulated is the AN/APQ-28 Search Radar being designed for use on the P7K aircraft which are scheduled for delivery to the fleet in the very near future. The AN/APQ-28 radar is known to be a complex unit requiring special skills for its operation which must be developed as soon as possible. Therefore the project engineer can conclude that the RLS is required as soon as possible and that delivery is of prime importance.

In addition to the above, the project engineer might note that the end of the fiscal year is approaching. In government procurements, money generally has to be obligated prior to the end of the fiscal year. Therefore, heavy pressure would be imposed on the procuring agency to enter into a contract as soon as possible. In addition to the fiscal year deadline, the buyer is limited by the amount of money that is budgeted for the procurement. For government procurements, budget information is restricted, but for commercial procurements such information is occasionally revealed and can serve as a guide to preparing the proposal bid.

Another factor of importance is that the requirement for the design approach is undetermined. Unless there are indications to the contrary, it can be assumed that the detailed design approach is left to the option of the offeror. The fact that the procurement is based on a performance specification would support the notion. In such a situation, project engineers should be prepared to elaborate to any degree desired on the advantages of their proposed design approach.

If there are indications derived from contacts with the technical personnel of the procuring agency that a particular design approach is viewed with favor, project engineers should be prepared to emphasize the advantages of their system if another design is favored or to reinforce the favorable views of the procuring engineers if their system is favored.

Another factor to use to advantage relates to the identity of the competitors who are trying for the contract award—what their probable approaches are, their probable costs, and their past performance, as well as the general attitude of the procuring personnel regarding each of the competing companies. Again, this type of information must be derived from experience and familiarity with the companies in the industry. The identity of the competing companies can usually be obtained by observing who the participants of the bidders' conference are or by deducing who the logical competitors would be. The design approaches of competitors can be deduced from past associations and conversations with personnel of the companies. It should be recognized that the fact that two companies are competitors does not preclude their associating with each other. In any general field of industry, there is a constant shifting of personnel among companies, and with the shifting, there are always exchanges of information which the new employer company records. Thus, Acmen Electronics can have a reasonable idea of the technical approaches that might be proposed by its competitors.

All the intelligence type of information can be correlated and organized to hypothetically establish what might be the customer's attitude toward a design that might be anticipated as being offered by a competitor. Thus, assume that Acmen Electronics analyzes the customer's attitude of Company CDE in the following manner:

Design Approach: d-c analog using a single Gray Scale Transparency scaled to 1,000,000:1. (This approach is known to be under development.) Unstable and too large with limited resolution capability.

Price: Estimated about equal to the below cost of Acmen Electronics

Delivery: Estimated to require 18 months

Past Performance: Poor

Having deducted the above, the project engineer should stress the fact that the Acmen Electronics' design approach using an a-c analog system is very stable, that its 5,000,000:1 scale offers a compact data storage system, and that the performance of Acmen has been good in the past.

In essence, it is incumbent upon the project engineer to analyze the probable attitude of the buyer toward the competitors and to emphasize during negotiations the advantages that the Acmen proposal offers over the competitive proposals without directly mentioning the competition.

9.5 Knowledge of Product

It is mandatory that project engineers be intimately familiar with the technical and cost details of the items under procurement and be prepared to discuss these points with confidence. They could have at their disposal selected personnel who are expert in the different important areas. If the negotiations require the inputs of one of the experts of the negotiation team, the project engineer would act as the spokesperson after privately discussing the particular point under consideration.

In addition to the knowledge relating to the technical aspects of the landmass simulator, the project engineer must be able to discuss and answer questions relating to the construction, type of materials, and components to be used, and other related areas of the simulator.

The type of materials and components to be used in the simulator would be of concern to the user. For instance, since the simulator must be capable of uninterrupted operation, the project engineer must be able to describe how the materials and components to be used will serve to meet the reliability requirements. The optical system, to point out an example, requires a somewhat delicate alignment, and the design of the alignment mechanism would be of great interest to the user.

An area of the procurement which has been touched upon lightly but which is important to the user relates to the side items. The side items for the simulator procurement are the reports (item 2), drawings (item 3), and manuals (item 4). The manuals in particular constitute a troublesome area to most users and as a result, a considerable amount of discussion as to how the manuals are to be prepared, who will prepare them, what their content will be, etc., is held. The project engineer should be briefed in this area and be prepared to discuss the side items in sufficient detail to satisfy the prospective user.

9.6 Knowledge of Cost Figures

The cost of a procurement is generally of primary concern to a prospective user, and the negotiations will be concerned with cost figures to a major degree. The amount of detailed discussion and the extent to which a cost breakdown is explored are dependent on the type of contract that is solicited. In a CPFF type of contract in which the contractor is reimbursed for all allowed incurred costs, the direct cost proposed is more or less academic and a matter of audit. The pertinent issue is an evaluation of how efficiently a particular company can perform. For instance, if company A bid $100,000 on a job but past experience indicates a probable cost overrun of $75,000, then it would be judicious to make an award to company B, which bid $125,000 but is evaluated from past experience as having a realistic price. Thus, in a CPFF procurement, the issue of direct cost is secondary. The detailed cost negotiations will generally be concerned primarily with overhead rates and profit.

On the other hand, the cost negotiations on fixed-price types of procurement can be expected to include in detail how cost figures were derived and a justification for the extent of effort in any particular area. Project engineers must be cognizant of the extent of effort, type of effort, contingency factors for risk, spoilage, material costs, etc., in all the areas of the project. They should be familiar with the type of cost breakdown presentation shown in Table 6.1, Detailed Direct Charges: Shadow Computer. The information in Table 6.1 might be broken down further to show, for instance, that in the Shadow Stop Circuit the engineering hours are broken down into the following:

Electrical engineering	80
Design implementation	70
Learning (new employee)	20
Contingency	30
Total	200

Thus, in the negotiations, if the effort quoted for the Shadow Computer is under attack, the project engineer could concede the learning effort (twenty hours) and the contingency effort (thirty hours) as a trade-off for some other consideration. In like manner, the project engineer knows that the time required for the engineering and design effort cannot be reduced without placing the company in a position that would result in a loss of contract.

Other areas of negotiation would involve labor rates, overhead rates, and G&A rates which essentially become a matter of audit. The establishment of labor rates is a matter of record. The overhead and G&A rates are derived by a formula based on statistical and cost accounting data as verified by government auditors or commercial accounting firms.

9.7   Negotiating Specific Contract Clauses

Innumerable variations of contract terms exist. Any procurement organization adopts a group of contract clauses which constitute procurement policy statements of the organization and are applicable to practically all contracts. These standard contract terms referred to as contract boiler plate are generally not negotiable by either party. Since most of the boiler plate terms are commonsense requirements and are characteristic of any business agreement, the offeror would have little reason to try to effect any change in them.

However, the terms of special clauses applicable specifically to the procurement under consideration are another matter, and the obligation that such clauses might place on the offeror can very well determine whether the contract will be profitable to the contractor or whether their presence in the contract imposes such a burden on the contractor that the company would lose money in meeting the contract requirements.

Many of the special contract clauses are intended to clarify which of the contracting parties has the obligation to perform in a particular area. In the case of the landmass simulator, one area of clarification is to establish which party is to be responsible for securing data that would be necessary for simulating the characteristics of the AN/APQ-28 radar. This information must be obtained from the manufacturer of the radar, known to be the Dalar Manufacturing Company of New York, and various government agencies who are concerned with the procurement of the equipment. Since the radar is still in the early stages of production, the data may not be formally documented and probably exist as preliminary drawings, manuscript documents, and other rough forms. Since the effort necessary to obtain the data could be significant, the establishment of the contractual obligation to perform this task is necessary and would be a point of negotiation. If the Acmen Electronics Company is to assume this obligation, then the contract should specifically establish this responsibility, and Acmen should negotiate adequate funds and time to carry out this responsibility.

In negotiating a contract, project engineers must be acutely aware of the implications that are involved in any special contract clause, and they must establish the minimum acceptable consideration for assuming the contractual responsibility in any area.

9.8 Negotiating Penalty Clauses and Ceiling Contract Price

Penalty clauses relate generally to delivery wherein a specified amount of money is withheld from the contractor for each day of delay in delivery. A delivery incentive clause is usually associated with a penalty clause whereby a contractor is rewarded for early delivery on the basis of a specified sum of money for each day that the date of the delivery is bettered.

The project engineer must be able to evaluate the penalty and incentive figures against the possible delays that might be experienced or against the improved delivery that might be possible. The actual figures can be readily obtained from those derived in the PERT scheduling discussed in Chapter 11. The optimistic, most likely, and pessimistic estimates of times calculated for the various activities could be used as a basis for establishing the range of expected delivery dates and thereby could be used as the schedule limits by the project engineer during negotiations.

If, for instance, a penalty figure of $100 per day is being considered for delivery beyond the target delivery date and the most pessimistic delivery derived from the PERT calculations is sixty days, then the project enginer must be prepared to consider a $6,000 possible maximum penalty. Then this penalty amount must be evaluated against the incentive, which might be $200 a day for every day that delivery is made ahead of the scheduled date. If the PERT calculations indicate that the most optimistic delivery is twenty days ahead of target date, the maximum possible incentive is $3,000. The project engineer must thus strive to negotiate the highest possible incentive rate and the lowest penalty rate.

The target cost is based essentially on the cost estimate as derived in the procedures as suggested in Chapter 6. It represents the best effort on the part of the contractor.

In negotiations involving FPI contracts, one very important factor is the amount of the ceiling cost of the contract. The offering of any company such as the Acmen Electronics Company will include a ceiling figure based essentially on the risk factors inherent in the procurement. The ceiling figure would be a negotiation factor upward or downward depending on what concessions the project engineer must give or receive during the course of the negotiations. If, for instance, the negotiations resulted in a reduction of the estimated engineering hours required and thereby in a reduction of the target cost, then the project engineer should counter with a request for a higher ceiling figure for the contract when that phase of the negotiation is reached.

Negotiations involve a large number of variables which project engineers must be able to juggle on the spot. Therefore, the more negotiation tools and guides they possess, the less chance they have of making concessions which might result in an unattractive contract for their company.

9.9 Negotiation Tactics

The first thing to establish at the negotiation table is that the negotiators for the customers have the authority to make commitments for their employers. If this is not established and agreed on, the discussions will merely be an exercise, and project engineers should conduct themselves accordingly. In other words, project engineers should not entertain making any concessions if customers will not be in a position to make concessions.

However, assuming that this basic prerequisite is met, the discussions can be initiated. Probably the most important single fundamental rule is to let the other party do the talking to the greatest extent possible. There is a certain amount of psychology in this approach. When the other party is allowed to dominate the conversation, it is satisfying its ego and thereby would be more prone to agree to an easier bargain. Then, too, the party doing the talking will often reveal its position. The astute project engineer can thereby use the information thus obtained advantageously.

As previously mentioned, project engineers should insist beforehand that they are acting as the spokesperson for their team. This approach will rule out the possibility of any other person's inadvertently saying something that might weaken the position of the project engineer.

There are numerous other fundamentals that should be noted in discussing tactics for negotiating. One is the element of timing as far as making concessions is concerned. Concessions should be held back for the time when maximum benefit might be derived. Also, points to be made in favor of the offerors or against some requirements which the opponent is attempting to establish should be timed to achieve the greatest favorable impact.

Project engineers should camouflage their real objectives to the greatest possible extent if the objective is of vital concern to them. For instance, suppose that the accuracy of the position of the simulated aircraft imposes a very serious design problem for Acmen Electronics. Instead of revealing this particular point, the project engineer might guide the discussion toward the number of engineering hours estimated in the area in question and be ready to concede a reduction in this area in return for a relaxation of the specified tolerances without making the discussion about accuracy an issue or jeopardizing the procurement.

Since negotiations are in effect a battle of wits, the use of subterfuge is not considered unethical. Because of the general acceptance of this idea, the use of tactics by which one of the negotiating parties might deliberately try to confuse an issue in order to achieve some objective is a common occurrence. Some of the tactics used in confusing the opposition are introducing trivial points to divert attention from some weak or vulnerable point in a proposal, raising a myriad of questions to put the opponent on the defensive, and guiding the opponent's questions to focus on the offeror's strong points.

During the process of negotiation in which each side is striving to gain an advantage at the expense of the other party, pressures and tensions can easily mount and emotional outbursts of one form or another can easily occur. Negotiators who lose control in such a manner can generally be assumed to have weakened their position and, in many cases, to have lost their negotiation objectives. Thus, a fundamental requisite of negotiators or project engineers is the ability to detach themselves from the issues of the procurement and pursue their aims in a completely objective manner.

As far as the tactics and conduct of a negotiation are concerned, project engineers should always keep in mind and recognize that their adversaries have a responsibility to protect the interests of their employers.

Since project engineers will undoubtedly become involved in future negotiations regardless of the outcome of the one at hand, they must always conduct themselves and their tactics in such a manner so as to command the respect of the opposition.

9.10   Qualities of a Negotiator

Because of the importance of the negotiation phase of a procurement, and because the success of a company in getting a contract award may hinge in large part on how the negotiation is conducted, it is felt that a discussion of the qualities that a negotiator must possess is in order.

A negotiation, by its nature, consists primarily of verbal communication between the two negotiating parties. Thus, the first quality that project engineers must possess is the ability to express themselves and their arguments effectively. There is no implication that these individuals must be silver-tongued orators, but they must possess the facility to translate technical, financial, or other information into clearly understandable language and express their points verbally.

Associated with the ability of expression is the facility to think clearly and rapidly. Upon entering a negotiation, the project engineer does not know what stand the procurement negotiator might take or what counterproposal might be made. Therefore, since the negotiation may take a sudden and unexpected turn, the project engineer must be able to appraise the implications of the turn of events and how the proposal is affected and then establish a course of action or a counterproposal. This must be accomplished while the negotiations are progressing with little time for consultation, review of records, or extensive analysis. In other words, project engineers must have the mental agility to evaluate the situation and decide a course of action.

Another attribute project engineers must possess is the ability to be objective and impersonal in their discussions. The opponent may have a valid point, and project engineers should be sufficiently objective to recognize and appreciate the opposing viewpoint. In conjunction with this quality, they must also assume a completely impersonal attachment to their proposal so that any deprecating remarks or criticism will not be taken as a personal criticism.

Patience is a virtue under any circumstances but is particularly important for a negotiator. The ability to subdue one's impulse to speak up when the opposition is attacking one's proposal, especially if the basis of the attack is erroneous, requires a large amount of restraint and patience. However, the patient negotiator should keep in mind that there will be an opportunity for rebuttal and more important should realize that in the very act of speaking, the opposition is exposing its position, giving forth valuable information, and in effect weakening its position. Therefore, at every opportunity, the project engineer should encourage the opposition to speak up and should certainly not interrupt to throttle any talks being made by the opposition negotiator.

9.11 Summary

Negotiations are conducted to clarify any areas or questions relating to the technical approaches, cost figures, or other areas which relate to the procurement. Project engineers, in serving as the key negotiator, strive to counter any objections to their proposal, impress the buyer with the merits of their offer, and obtain a contract award at the most favorable terms and price for their company.

Prior to engaging in a negotiation, project engineers must make a thorough analysis of the technical approaches and cost breakdown of their offering and must be adequately prepared to talk with confidence and conviction on any area. Although they should form a team of individual experts in key areas and have their team participate in the negotiation, they should act as the spokesperson and have complete control over what is said or presented.

Because of the many variables that exist in any procurement, all of which are important, project engineers should arm themselves with charts or other readily interpreted references which indicate the boundaries of the variables beyond which they cannot go. The three major variables which must be weighed against each other are price, technical requirements, and delivery schedule.

An evaluation and appreciation of the buyer's position should be based on the best available intelligence in order to serve as a guide for the negotiations.

Of particular importance in any negotiation is the effect that special contract clauses would have on a procurement. The project engineer must have sufficient appreciation of the procurement to evaluate the impact of any special contract clause that might be proposed. In general, the project engineer should avoid assuming any obligation without adequate contractual consideration.

Negotiation is a complex art in which each party seeks to gain a benefit at the expense of the other. Project engineers must be adept at using different tactics for accomplishing their objective. The nature of the conduct of negotiations demands that project engineers be able to express themselves clearly and logically and be able to comprehend quickly the various complexities that may develop so that they can modify their proposal as required.

Chapter Ten

The Legal World
of Project Engineers

10.1 Contractual and Legal Problems

Project engineers and the driver of the family automobile have one thing in common. They both have been given a certain amount of authority, but errors in exercising such authority can result in penalties. The driver of the automobile, though licensed, can be subject to penalties because of damage resulting from negligence. Project engineers, by unauthorized direction or failure to act, can violate contractual agreements which would lead to penalties for their organization. Ignorance of the regulations and law in each example will not provide relief from the damages that might be experienced.

In order to avoid legal and contractual complications that may arise from their actions or inactions, project engineers must have a working knowledge of contract law and develop a sensitivity toward actions that might lead to contractual complications. In addition, project engineers must know all the special provisions of their contract, the detailed requirements of the specification and schedule, and the type of procurement that is involved so that they can direct their project with the confidence that their actions are legally and contractually correct.

From the project engineer's point of view, the rules that regulate the solicitation and proposal evaluations prior to the award of a contract are different from that of a procurement involving commercial parties. The federal, state, city or other government agencies take the position that the public funds that are used for a procurement belong to every tax-payer—including the companies seeking a contract. Therefore, specific rules and regulations which are designed to provide completely equitable treatment to all qualified companies and ultimately to the contractor govern such procurements.

Procurement actions prior to the execution of legal contracts which involve commercial firms are not subject to the type of regulations that apply to government procurements. The general philosophy is that organizations are free to spend their own money as they wish. Customers are usually not bound by specific regulations when conducting solicitations and may generally award to whomever they wish on whatever basis they desire. In order to discourage unfair practices in solicitations involving private funds, many states have adopted commercial codes which set standards for the conduct of advertised solicitations. However, because of broadness of commercial codes, companies wishing to participate in such procurements must make their own judgments as to whether they will be treated fairly and whether the contract award will be made on an impartial basis.

Regardless of whether a solicitation for a procurement by the government or private company is involved, it is incumbent on project engineers employed by offerors to make a thorough appraisal of their company's chances of success in a procurement prior to the deciding if money and effort to be spent on a proposal is worth the risks of not being successful of an award.

The different environment for the government and the private industry procurement continues prior to and after the signing of the contract. In both cases, the terms of the contract are dominant in establishing the responsibilities of the contracting parties. However, when public funds are used, special rules and regulations of the government agencies (federal, state, city, etc.) place far greater restrictions on what terms can be put into a contract and what either party, particularly the procuring agency, can do.

When contractual problems develop, all attempts to seek a resolution should be made by the two parties. The engineers of the contracting parties would have a major responsibility of protecting the interests of their organizations and at the same time seeking an equitable solution. If a resolution cannot be reached on a government procurement, the contracting officer can unilaterally direct the contractor to proceed on a particular course. Contractors must proceed as directed but then have the option to formally appeal the direction and prepare a claim against the procuring activity for directing that they pursue a course which is not in accordance with the terms of the contract.

In pursuing a claim, the contractor's argument would be presented to government agencies who sit in judgment of contractor's claim. Where the contractor has exhausted the channels of appeal provided by the government agencies, the case can be thrown into civil courts. Judgment will then be based upon the same laws and criteria that would be used for contract disputes between private parties.

For contracts involving private parties, the resolution of contractual problems may be more complex. Often, a performance bond is involved which protects the customer from damages due to work interruptions, and the issues would be referred to the legal courts for prompt resolution.

Figure 10.1 shows the two parallel paths that would be followed in the execution and appeal proceedings of a government and a private industry procurement. The ultimate judgment in either type of contract rests with the civil courts of law as noted in Figure 10.1. What is inferred is that philosophically, the legal aspects of the regulations that govern the conduct of procurements by government agencies must ultimately be able to be supported by the common law of the jurisdictional courts.

Since a contract normally involves a legal agreement between two parties, the contractual pitfalls exist on both sides of the fence. Incorrect

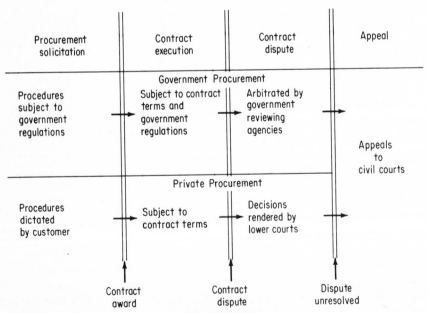

FIG. 10.1 Legal paths for government and commercial contracts.

actions by the project engineer for the customer can be as costly to the buyer as errors made by the contractor's project engineer. Practically any facet of a contract or any action taken during the planning and execution of a contract can involve legal and contractual complications. The subject is too broad to cover in detail except to identify the most common problem areas in which the project engineer would become involved.

The discussions in this text will relate primarily to the legal and contractual world of government procurements since government agencies are the customers for most large contractors handled by project engineers.

10.2 Specifications

The writing of the specification is usually accomplished under the direction of the customer's project engineer. As was discussed, the design specification describes not only what is required but how the equipment is to be designed and built. The performance specification is generally limited to describing what is required so that the details of how the equipment is to be designed and built are left up to the contractor. In either case, the customer's project engineer must exercise caution to avoid contractual complications in the writing of the specification. Some of the major specification flaws that might lead to difficulties are as follows:

1. *Omissions:* Whatever is required should be fully described. If a specification contains a void, the contractor could provide a component, system, or design that may not represent what the customer desired; but because of the specification omission, the customer may have no option but to accept what the contractor provides.

2. *Nebulous requirements:* A specification requirement that can be interpreted in more than one way could result in having the contractor provide for the more simple and less costly requirement—generally to the detriment of the customer.

3. *Unclear accuracies:* Accuracies are usually expressed in percentages. Disputes can arise if the base for the percentages is not clearly cited. It is essential that accuracies be clearly defined for the entire ranges of operation and that care be taken so that even farfetched interpretations cannot be made from the terms of the specification.

4. *Inconsistencies:* The details of a specification must be consistent throughout. If in one part of the specification a requirement for a performance parameter is less stringent than the same requirement expressed in another part of the specification, the contractor will normally provide the less demanding requirement. Often, what the contractor indicates will be provided will not be satisfactory to the user, but contractu-

ally, the customer has no basis for demanding what was needed. The result is that a contract amendment, involving additional funds, has to be negotiated in order to accommodate the customer's requirement.

5. *Impossible requirements:* A requirement may have been expressed in the specification or the schedule which, as the project progressed, proved to be impossible to satisfy. A whole battery of legal questions will surface in any such situation, but the one dominant question raised would relate to determining who was the more knowledgeable party at the time of contract. In the absence of any other overriding circumstances, the customer is usually considered to be the more knowledgeable and therefore it has been held that in such circumstances the contractor could not be held responsible for not meeting that particular specification requirement. The point to be made is that if a project engineer suspects a requirement to be impossible to achieve, the information should be brought to the surface for discussion and resolution prior to contract award.

6. *Deficient compliance:* Contractors may deliver equipment which falls short of meeting all the requirements of the specification. The courts have taken the position that if contractors are in substantial compliance with the specification or delivery requirements, the contract cannot be defaulted. However, even though contractors are not held in default, they are not relieved from submitting to an adjustment in contract price to compensate the customer for receiving less than what was contracted. It is incumbent on the contractor's project engineer to assure that all requirements of the contract are satisfied since the above noted adjustments could very easily result in a new contract price which represents a loss to the company.

10.3   Proposals and Evaluations

The proposal submitted for a project is the responsibility of the offeror's project engineer and the evaluation of proposals that are received is the responsibility of the customer's project engineer. In order to assure maximum effectiveness of the proposal and realize the greatest economy of time and money, the proposal should avoid addressing anything but that which is identified in the TPR. Since offerors are in competition with each other for the contract award, a proposal should avoid unnecessary embellishments that would inflate the proposed cost. In other words, proposal design described by the project engineer should provide no more than that which is called for in the specification and contract schedule.

The preparation of a technical proposal for a major procurement can be a costly effort for each offeror. Because of cost, most companies may decline to participate in a procurement unless their analysis of the procurement leads the company to conclude that its chance for award is worth the cost involved. The participation decision was discussed in an earlier part of the text. Because of the expense for creating proposals, the government and, in many instances, the larger private customers will purchase technical proposals from qualified offerors, thereby assuring adequate participation.

The offeror's project engineer should structure the proposal to satisfy the criteria established by the user. To offer a design that may exceed the minimum requirements expressed in the specification would be advantageous only if offerors do not penalize themselves because of higher cost. Contract law for government procurements generally establishes that the offeror satisfying the minimum requirements of a specification and having the lowest price would be awarded the contract. If an offeror proposes a unique or improved capability that the customer determines is desirable even though more costly, the customer, particularly if it is the government, will have to amend the solicitation to permit all offerors the opportunity of proposing on the improved capability.

If, with its own resources, a company had developed a unique design, the company can claim proprietary rights. In cases in which the product of a development is patentable, the company would have privileges of selling the product to a government agency or private party on an exclusive basis. If the unique design is not patentable, the company can claim proprietary rights on government procurements and enjoy a position of a sole source contractor. When dealing with a private customer, the company would handle the product design as a trade secret and thereby enjoy a position as the exclusive source for the product.

In executing a contract, companies frequently create a design which is unique and could be patentable or at least be handled as a trade secret. Because the development was the result of effort funded by the government, the company would enjoy exclusive rights only for procurements involving other private organizations. For government procurements the development is considered property of the government and the design information would be made available to all offerors.

Clarification discussions are often conducted as part of the procurement process. There are two basic types of clarifications: those questions raised by offerors relating to the specification or contract schedule prior to submission of the proposals, and those raised by the customer regarding some point in a proposal. One basic procurement principle is to keep all offerors apprised of all information disseminated by the customer. If one offeror solicits and receives an answer to some question on a procurement, government contract regulations require that all offerors be made privy to the questions that were raised and the answers that were provided. The same principle applies to data, information, and other pertinent intelligence.

The technical evaluation of the proposals almost always involves challenges from the unsuccessful offerors. Generally speaking, the courts of law and the government agencies reviewing protests based on the technical evaluations of proposals have avoided involvement where issues of judgment of professionals are involved. The most common issue of this type relates to the technical judgment of engineers. The courts and reviewing authorities in such cases have generally become involved only in questions relating to procedures.

In commercial procurements, the procedures and rationale for evaluation are established at the time of solicitation and will vary from case to case. Legal confrontation can arise when the customer violates ground rules that may have been agreed upon among the customer and offerors since such violations in themselves may have been a breach of understanding. However, such legal cases are rare. The desire of a customer to preserve its good reputation in procurements is probably more of a deterrent to unfair practices than the threat of possible legal actions.

When a procurement involves public funds, the regulations play a dominant role for maintaining proper procurement procedures. The major points that the project engineer evaluating proposals for a government procurement must comply with are as follows:

1. The proposals must be evaluated in the areas cited in the TPR document.

2. The criteria requirements for equipment performance which the proposed design is to satisfy must be consistent with the requirements cited in the specification.

3. The TPR document must include information relating to the evaluation criteria and the factors that will be considered for making a contract award.

4. All proposals must be evaluated in the same areas and against uniform criteria.

10.4 The Contract Document

A legal contract comprises two basic elements. There must be an agreement or meeting of the minds between the contracting parties and there must be an exchange of consideration. In the case of a procurement, the consideration would be the goods and/or services the contractor agrees to supply in return for money. The language in the various documents that comprise the contract constitutes the agreement which is attested to by the signatures of the parties representing the contractor and the customer. During technical clarifications and negotiations various points of the specification, schedule, and other documents are usually discussed and clarifying language established. It is essential that the project engineer make certain that the clarifications established are reflected as revisions to the documents that will comprise the formal contract. Failure to document agreed-upon clarification into the contract prior to its execution can result in complications later on because of the same issues being raised during the course of implementing the contract.

Once a contract is executed by the parties involved, the courts would have ultimate jurisdiction of formal agreements regardless of whether the contract is between private organizations or whether public funds of a government agency are involved. It is incumbent upon the project engineers of the parties to know all the details of the contract in order to avoid placing their company or organization in the vulnerable position of breach of contract.

10.5 Execution of the Contract

The scope of contract law is broad and represents a specialized legal field. As noted earlier, project engineers rarely have any formal legal training; but since they have to survive in a legal environment, they must be knowledgeable of the major ground rules of contract law that relate to their project.

The following are some legal points that project engineers must recognize during the execution of the contract:

1. *Constructive changes:* The terms of the original contract rarely remain unchanged during its life cycle. Discussions between the project engineers and other officials of the contracting parties will take place. Some of the discussions could relate to points that represent changes to the contract terms. Any discussion regarding a possible change to the contract should address the impacts of cost, delivery, equipment performance, etc., in depth. If a change is desired, such change shall be reflected as a revision to the specification or schedule and treated as a formal contract change.

However, if as a result of any discussion the contractor acts on a project engineer's direction, suggestion, or some other type of expression, a construction change to the contract may have been precipitated by the project engineer. Such constructive changes usually occur when there is no clear understanding between the contracting parties that any change must be supported by an official documented request or direction. The courts have even stretched the point and ruled that even silence on the part of a project engineer with regard to some point that was made by the other party inferred agreement and therefore became a constructive change.

Constructive changes can and have resulted in significant changes to the cost, delivery, equipment characteristics, etc., of a contract which were not favorable to the project engineer charged with the constructive change responsibility. The point is that project engineers must be very explicit in all their discussions with the other contracting party to avoid having any of their actions result in an unintentional change to the contract.

2. *Termination:* Practically all contracts contain clauses pertaining to the circumstances that may cause the termination of a contract and the legal consequences resulting from such action. The principles relating to contract termination for government and private industry contracts are similar in nature.

Customers can usually terminate "for their convenience," but in so doing are liable for all expenditures incurred by the contractor including profit and termination charges. Obviously, a convenience termination is an expensive luxury for the customer and would be used only in those rare cases in which unforeseen circumstances of need or money dictated the termination action. The role of project engineers in a convenience termination action would primarily involve the establishment of the rationale for negotiating the termination settlement which best serves their employer's interests.

A termination by default action occurs when contractors are unable to complete their contractual obligations or when there is substantial evidence that insufficient progress is being made in prosecuting the contract. Generally, termination by default actions are not valid when circumstances beyond the control of the contractor or "Acts of God" are responsible for the situation.

When circumstances within the control of the contractor such as poor management, etc., cause the contract effort to come to a halt, the justification for termination by default is clear-cut.

However, when default termination is undertaken based on the contention that the contractor's progress in prosecuting the contract is unacceptable, the issue can be controversial. For such action, the project engineer for the buyer, who is initiating the termination action, must be prepared to prove to the reviewing authorities and court that the contractor is faced with such insurmountable problems that one or more of the significant objectives of the contract cannot be accomplished. Examples of such problems might include one of the following:

1. *Technical difficulty:* The design approach adopted by the contractor has lead to problems which the customer's project engineer judges cannot be solved within a reasonable time. Because of cost considerations, the contractor may not be willing or able to change the design

approach. The customer may determine that the only option left is to terminate for default.

2. *Financial problems:* Because of money problems, the contractor is unable to perform satisfactorily. If no financial relief is available or bankruptcy is possible, termination action would be warranted.

When a convenience termination is executed, the customer is liable for all costs incurred by the contractor but takes title to anything purchased or produced. In a default termination, the contractor is reimbursed for only those items that may have been accepted. Such items often involve only reports, drawings, or other minor items. Contractors not only have to absorb all incurred costs for undelivered items but are liable for the difference in their contract price and the possible higher price that an alternate contractor may charge to provide the equipment or items originally intended. It is the prime responsibility of the contractor's project engineer to manage the project so that a default termination is avoided. In the final analysis, the project engineer must bear the major responsibility for such a disastrous eventuality.

3. *Data and/or furnished equipment:* A contract may require that the customer deliver or make available specific parts or equipment to be used for the product to be delivered or data required for the design of the contracted item. It is the responsibility of the customer's project engineer to not only have the data or furnished equipment delivered to the contractor on schedule but assure that what is required is complete and adequate for use on the contract. If, for instance, engineering drawings that the customer is obliged to furnish are inaccurate, incomplete, or not delivered as scheduled, the contractor has a legitimate basis for a claim against the customer. It is the responsibility of the customer's project engineer to be aware of the legal and contractual implication that the customer has assumed and exert all efforts to live up to the customer's side of the contract.

4. *Progress payments:* Most large contracts which require a long period of time to execute provide for progress payments to be made in accordance with some agreed-upon performance criteria. One common criterion or milestone is the completion of the engineering effort, which is the point in the project life when a certain agreed-upon percentage of the contract price will be paid to the contractor.

It is vital to the project engineer to know in detail the content of the progress payment contract terms and manage the project to assure that the progress payment milestones are achieved on schedule. The following are the major reasons why progress payment terms are important:

a. Cash flow enables the contractor to meet current expenses thereby avoiding the necessity of borrowing money at high interest rates.

b. Meeting scheduled milestone dates is evidence that the project is not experiencing problems. A project that falls behind its projected schedule will also exceed its projected cost—which usually must be absorbed by the contractor.

5. *Penalties, liquidated damages, and incentives:* Delivery schedules on contracts are always important. Often a contract contains terms which penalize late deliveries or provides for a reward for deliveries that are earlier than specified. A contract with penalty clauses imposes major responsibilities on both the project engineer for the contractor as well as the customer. The contractor must exert every effort to live up to the terms of the contract and in particular to make special efforts to avoid situations that would lead to penalties or damages. The customer's project engineer must recognize and live up to any obligations that the contract imposes. If for instance, the customer fails to deliver data or equipment to the contractor as required, the contractor could claim that its failure to satisfy, covered by penalties or liquidated damage clauses, is due to the fault of the customer and therefore the imposing of penalties is not valid. Case histories have confirmed that the position noted is valid and contractors might thus be relieved of their responsibilities in the area concerned.

10.6 Summary

The regulations for a program using public funds such as federal, state, city, etc., procurements differ from the rules that might apply to procurements between private organizations. However, protests that are pursued to the ultimate end would eventually fall under the same legal consideration in the courts of law.

The actions of project engineers can contractually commit the organization or activity they represent, therefore it is essential that they know the terms and conditions of the contract and be aware of the legal consequences of any of their actions or lack of action.

Chapter Eleven

PERT and Other
Project Management Tools

11.1 PERT (Program Evaluation Review
Technique) Concepts

The greatly increased complexity, size, and amount of development required by modern commercial as well as military systems have made obsolete the traditional managerial methods and controls for estimating, scheduling, and cost control. In seeking more effective means for managing complex procurements, the PERT and other management systems were developed.

The PERT system is a managerial tool for determining at any point in the life of a program precisely what the status of the program is and where the trouble areas lie.

Its greatest value is that it signals management in advance when any difficulties develop in any specific area which will adversely affect the planned program schedule or budget.

The basic concept of PERT is that the program is divided into discrete detailed scheduled tasks which are drawn up into an integrated network. All the significant variables of time, resources, and technical performance are allocated to each task or activity. A system of systematic reporting is then implemented, which enables management to compare actual performance with the original program plan, thereby permitting a continuous check on the program status.

As a management tool, PERT enables the project engineer to shift resources from noncritical to critical activities, thereby permitting the concentration of resources in those areas that were signaled as experiencing difficulties.

11.2   PERT Definitions

The PERT system uses a unique language. The following are the most fundamental terms which are used:

1. *Activity:* An element of work effort in a program.

2. *Event:* A specific point in the program usually representing the start or completion of an activity. An event does not have any dimension in time or effort.

3. *Network:* A graphic representation of a program consisting of activities and events which are shown as interconnected paths.

4. *Most Likely Time, m:* The most realistic estimate of time that it would take to complete an activity.

5. *Optimistic Time, a:* The shortest period of time that the completion of an activity would consume.

6. *Pessimistic Time, b:* The longest period of time that the completion of an activity would consume.

7. *Expected Time, T_e:* The period of time that is predicted for completing an activity. The expected time is statistically derived from the most likely, optimistic, and pessimistic times as expressed in the formula

$$T_e = \frac{a + 4m + b}{6}$$

8. *Cumulative Expected Time, T_E:* The earliest date that can be anticipated for the completion of a specified work effort or efforts. T_E is the summation of all the expended time T_e along a particular path.

9. *Latest Allowable Date, T_L:* The latest date on which an event can occur without delaying the completion of the program. The latest allowable time is calculated by subtracting the expected elapsed periods or expected times (T_e) of activities from the date of the last event. If the end date coincides with the date represented by T_E, the $T_L = T_E$.

10. *Positive Slack Time:* The amount of excess time predicted for the achievement of a particular event. Negative slack indicates the amount of slippage that exists prior to reaching a particular event. Slack time is the difference between the latest allowable date and the expected date ($T_L - T_e$).

11. *Critical Path:* The path of a network that requires the longest

period of time to complete. It is the path that possesses the smallest positive slack or the greatest negative slack.

11.3   Operation of PERT

The operation of PERT can be divided into the following five broad categories: (1) establishment of objectives, (2) creation of plans, (3) establishment of schedules, (4) evaluation of performance, and (5) arrival at decision and action.

These categories comprise the PERT cycle which is illustrated in Figure 11.1. In addition to the logical flow of sequential information from one category to the next, the corrective feedback loop which constitutes one of the most valuable characteristics of PERT is shown. The corrective feedback permits the project engineer to implement changes in the program or plans of action on schedules if the program objective of schedule or cost is in danger of not being met. In addition to being an essential function in setting up a PERT plan for a program, the establishment of the objectives enables the project engineer to crystallize the project goal and document the project goals for management and other interested parties.

The RLS program involves several deliverable items in addition to the actual hardware. The other items indicated on the contract schedule are the reports, drawings, and manuals, and a comprehensive network for the program would take cognizance of all these deliverable items. Since this text will discuss only the principles of PERT, in the interest of simplicity the development of the network will be restricted to the hardware.

The PERT plan of any major item of a program such as the hardware requires that the effort be divided into subtasks, each of which comprises a definable area of work effort. Each of the subdivisions is identified as a work package and constitutes the effort required to complete a specific job. The work package would be represented as one or more activities on the network.

The creation of the plans involves the translation of the work packages

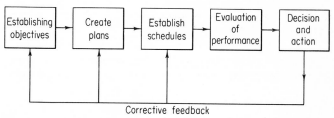

FIG. 11.1   The PERT cycle.

into activities and events which are graphically described as a network. In creating the network, cognizance must be given to the sequence in which each activity is performed and to the earlier events which must be reached before any particular activity starts.

When the activities are set forth, the expected time T_e is derived from estimates made for the optimistic, pessimistic, and most likely times. The project engineer would obtain these figures from the individuals who would be responsible for performing the work effort of the activity. The time estimate would be based on using available personnel and resources based on a 40-hour week.

The establishment of the schedule is the conversion of the network elapsed activity times into calendar dates. In deriving the schedule, the project engineer must take cognizance of the following basic factors:

1. The contract delivery date and the date on which work is to be initiated.

2. Available personnel and resources of the company (as contrasted with the personnel and resources available for any particular activity).

3. Constraints of different activities. A constraint is an activity that must be completed or an event that must be reached before the work effort of another activity can be initiated. Constraints often present critical areas in a PERT program since they constitute limitations in facilitating the meeting of schedules.

In drawing up the schedule of the program network, the project engineer will probably find that several revisions will have to be made before a PERT plan can be created which will reflect a program planned completion date which is consistent with the contract completion date. The trial and error involved in creating a consistent plan often may involve compromises with what may be the ideal program to the company. For instance, the original plan may have been based on purchasing a particular assembly as a subcontract item having a long lead time. By building the item in-house, the undesirable long lead time might be avoided even if the cost is higher. If the shorter schedule that can be realized by having the item produced in-house justifies the added cost, then the PERT network would be revised.

The evaluation of the performance and the decision and action phases of the PERT cycle is made by each level of management. The officials at each level would study the information derived from PERT from a different point of view and would implement some of the following actions within the framework of their authority: assessment of performance and action, execution of action, and transmitting necessary information about unresolved problems to the next higher level of management as required.

There are numerous types and forms of reports that have evolved from PERT programs. In general, the reports serve to advise management of the program schedule and cost status as of a specific date, a comparison of the actual program status with the program as planned, prediction of program schedule and costs, the identity of areas which are potential or actual sources of difficulty, and other pertinent information of this nature.

11.4   Implementing the PERT Plan

The implementation of the PERT plan will be illustrated for the RLS.

The project engineer, in setting up the plan for the RLS, would verify the work breakdown structure which was illustrated in Figure 3.1 in conjunction with establishment of the design detail discussion. The PERT can be based on practically any desired tier of design detail. The degree of detail of a network is proportional to its complexity so there is an optimum point of detail in a network beyond which the amount of monitoring effort is not proportional to the value of information that is derived. As far as the case of the simulator is concerned, the project engineer has determined that the tier III design detail is adequate for the PERT plan.

The next step that the project engineer would take is to divide each of the elements of the tier III design detail or work breakdown into work packages.

The Shadow Generating System of the landmass simulator is one of the work breakdown areas shown in figure 3.1 and will be discussed in deriving the PERT plan. The project engineer would divide the Shadow Generating System into the following work packages: (1) analysis, (2) electronic design, (3) mechanical design, (4) reliability engineering, (5) procurement of parts, (6) drafting, (7) manufacturing, and (8) test and checkout. It should be recognized that the Shadow Generating System is only one of the several subsystems that comprise the RLS. The time cycle to complete the Shadow Generating Subsystem is shown as 45 weeks and is not inconsistent with the time of 65 weeks shown in Figure 3.5 which is required for the complete simulator.

The next step that would be taken is to select the events. The selection is done in consultation with the individuals of the program team who are to be responsible for the completion of the various work packages. The events must represent specific beginnings or endings of effort in the program.

The PERT network is organized with each event numbered and connected to another event as shown in Figure 11.2. The arrows indicate the flow of work in a logical sequence. The solid arrows represent actual effort requiring the completion times shown by the groups of numbers associated with each arrow. The dotted arrows generally represent constraints representing zero time. For example, event 03, start design, cannot be started until the event 02, complete data gathering, is completed, even though activity 01-03 could be completed in a shorter time than activity 01-02 as shown in Figure 11.2.

The dotted arrows are also used to simplify the graphic network and permit the creation of network paths for different types of effort. For example, event 03, start design, involves three types of design: electronics, mechanical, and reliability engineering. These are shown as events 04, 05, and 06, connected by dotted lines (zero time) from event 03.

After the project engineer has designed the program PERT network, the next task is to establish the elapsed times required for completing each activity. The sources of this information are the group leader who will be responsible for the effort of the different activities. The three

TABLE 11.1 Summary of Activity Elapsed Times for the Shadow Generating System PERT Network

Activity	Identity	Elapsed times	T_e
01-02	Data gathering	8-10-12	10.0
01-03	Design analysis	9-11-12	11.8
02-03	Dummy line	0	0*
03-04	Dummy line	0	0
03-05	Dummy line	0	0
03-06	Dummy line	0	0
04-07	Dummy line	0	0
04-08	Electronic design	5-6-8	6.2
07-08	Breadboard analysis	4-5-8	5.3
05-09	Mechanical design	6-7-9	7.2
06-10	Reliability engineering	4-5-6	5.0
08-11	Dummy line	0	0
09-11	Dummy line	0	0
10-11	Dummy line	0	0
14-10	Dummy line	0	0*
11-12	Prepare orders	1-2-3	2.0
12-16	Purchasing	12-13-15	13.2
11-13	Integration design	8-10-12	12.0
13-14	Dummy line	0	0*
14-15	Drafting	6-7-9	7.2
14-16	Dummy line	2-2-5	2.5*
16-17	Manufacturing and assembly	5-6-9	6.3
17-18	Dummy line	0	0*
18-19	Test	2-2-5	2.5*

* Critical path.

estimates for each activity are the optimistic, most likely, and pessimistic times. These figures are usually indicated on the PERT network as illustrated in Figure 11.2. Activity 01-02, for instance, has elapsed times of 9, 11, and 12 weeks representing the three estimates.

The PERT network is based to a large extent on a statistical analysis. The three estimates for each activity are averaged with extra weight given to the most likely time to derive an estimated elapsed time T_e which is used in scheduling.

Table 11.1 lists the different estimated times for each activity of the network illustrated in Figure 11.2.

The PERT network is intimately correlated with the organization of a program, and specific responsibility for each activity must be established and implemented with adequate controls and lines of communication.

The following are some of the inherent characteristics of a PERT network which would be observed when designing a PERT network system:

1. Any particular activity must be completed prior to the occurrence of an event. In like manner, an activity cannot be initiated prior to the establishment of an event. For example, from Figure 11.2 the drafting activity cannot be initiated until the event complete system design (13) has been completed.

2. All activity paths must be complete and cannot be duplicated or represent alternatives.

3. Any particular event can only occur once.

4. Only one activity line can connect any two events.

11.5 Use of the PERT Network

Once the PERT network has been designed and the expected elapsed times of each activity calculated, the project engineer can initiate its use as a management tool. The status of each event will be of primary concern to management, and special attention will be focused on the events. An examination of the network of Figure 11.2 will reveal that all activity paths ultimately lead to the final event, but the total of elapsed time T_e will differ for different paths. One path will require the largest total elapsed time and represent the path that is critical to the PERT network. This critical path and the completion of the events on the critical path will receive the most management attention.

For the network in Figure 11.2, the critical path is indicated by the double slashes on each activity of the path. The PERT network gives an excellent overall view of the program and enables the project engineer to shift personnel and other resources from slack paths to critical paths in order to render aid in overcoming areas of difficulty, as was suggested previously.

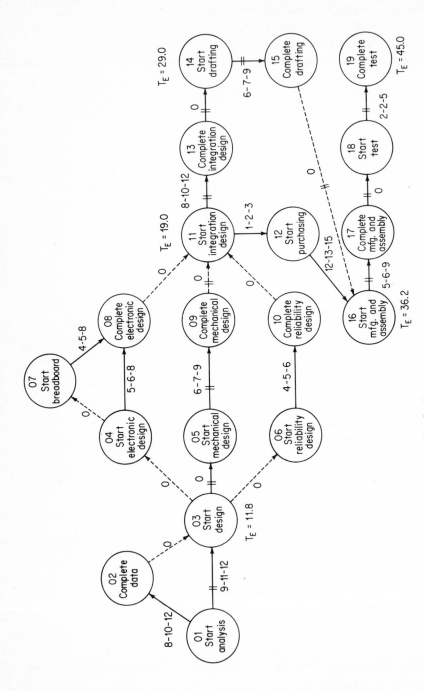

FIG. 11.2 PERT network for the Shadow Generating System.

**TABLE 11.2   Summary of Event Times and the Critical
Path for the Shadow Generating System**

Event	T_E	T_L	Slack	Critical path
01	0	0	0	X
02	1.0	1.8	0.8	
03	11.8	11.8	0	X
04	11.8	12.8	+1.0	
05	11.8	11.8	0	X
06	11.8	14.0	+2.2	
07	11.8	13.7	+1.9	
08	19.0	18.0	+1.0	
09	19.0	19.0	0	X
10	19.0	16.8	+2.2	
11	19.0	19.0	0	X
12	21.0	23.0	+2.0	
13	29.0	29.0	0	X
14	29.0	29.0	0	X
15	36.2	36.2	0	X
16	36.2	36.2	0	X
17	42.5	42.5	0	X
18	42.5	52.5	0	X
19	45.0	45.0	0	X

Other points to note in the PERT network are that different events will be reached at different times because of the variations in activity estimated elapsed times. The earliest time T_E in which each event can be reached and the latest times T_L that events must be reached without jeopardizing the project schedule are calculated in making the network analysis. The various values for T_L are obtained by working backward from the final event. The various values of T_E and T_L for all the events are shown in Table 11.2.

The slack for each event tabulated in Table 11.2 indicates the excess time available to complete the activities leading to a particular event. The slack time is calculated by subtracting the value of T_L from T_E.

Positive or zero slack time indicates that all events are at least expected to be on schedule and that no difficulties are expected. If the slack time for any event is a negative value, then the activities contributing to such negative slack times are experiencing difficulty and corrective action of some type is required.

11.6   Probability Features of PERT

One type of information that the PERT network reveals is the earliest time T_E that any event can be expected to be reached, which includes the final program event. The values of the various event earliest times are derived from the summation of the activity expected times T_e.

An examination of the activity elapsed times in Table 11.1 indicates that there are different spreads between the optimistic and pessimistic times; these spreads are a reflection of the degree of certainty of completing an activity within a specific time. For instance, if the project engineer were 100 percent certain that activity 01-02 would take exactly 11 weeks to complete, then the three elapsed times would be 11-11-11. The presence of no spread indicates maximum certainty of completing the activity in 11 weeks.

The graphic representation of the chances of completing an activity in any of the elapsed times can be shown as the distribution curve in Figure 11.3. If the spread of times is large, the curve would flatten out, and, conversely, a small spread would result in a narrow curve.

The probability of completing the activity in any time measured along the horizontal t axis is represented by the area under the curve at the particular point of interest. Thus, the probability of completing the activity within the optimistic time a is very small, and the probability of completing the activity within the pessimistic time b is very large (about 100 percent). The probability of completing the activity within the most likely time is 50 percent, which is the arbitrary chosen basis for the PERT statistics.

One important characteristic of the distribution curve is its standard deviation, which is a direct function of its spread. The standard deviation (SD) is measured from the left and right of the medium (M) along the horizontal axis. One SD designates points on the horizontal axis which are the boundary for 68 percent of the area under the distribution curve

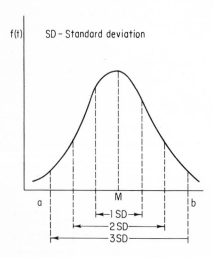

FIG. 11.3 Normal distribution curve typical of PERT activities.

TABLE 11.3   Summary of Calculations for Probability of Reaching Event 09 in 18.5 Weeks

Activity	Pessi-mistic time (b)	Optimistic time (a)	Spread (R) ($b - a$)	Activity SD = $R/6$	Variance (SD)^2
01-03	12	9	3	3/6 = 0.50	0.25
03-05	0	0			
05-09	9	6	3	3/6 = 0.50	0.25
				Total variance = 0.50	
				SD = 0.71	

$$Z = \frac{\text{scheduled time} - T_E}{\text{SD}} = \frac{18.5 - 19.0}{0.71} = -0.70$$

as illustrated in Figure 11.3. Two SDs designate 95 percent, and three SDs designate 99 percent of the area under the curve.

When a series of activities having different time distribution curves is dealt with, the SD of the curves can be correlated. Therefore a value identified as a variance is derived from the SD value by the formula

$$\text{Variance} = (\text{SD})^2$$

The preceding discussion relates to some basic statistical concepts of PERT, and a knowledge of these concepts is necessary for determining the probability of reaching any event in the program within a specific time. This is the type of information that the management would very often ask of the project engineer.

In Figure 11.2 and Table 11.2, it is noted that the earliest time T_E in which event 09 will be reached is 19.0 weeks. The time T_E is derived by adding the various values of T_e along the longest or critical path leading to the event. Based on the statistics of PERT, the probability of reaching event 09 in 19.0 weeks is 50 percent.

Assume that the project engineer was requested to calculate the probability of reaching event 09 in 18.5 weeks. The procedure for deriving such information is as follows:

1. Calculate the spread of each activity leading to event 09 (see Table 11.3).

2. Calculate the variance of each activity (Table 11.3).

3. Calculate the composite SD of all activities (Table 11.3). Composite SD is the square root of total variance.

4. Substitute values in the following formula to derive factor Z

$$Z = \frac{\text{scheduled time} - T_E}{\text{composite SD}}$$

TABLE 11.4 Normal Probability Distribution

Positive values		Negative values	
Z	Probability	Z	Probability
0.0	0.500	−0.0	0.500
0.1	0.540	−0.1	0.460
0.2	0.579	−0.2	0.421
0.3	0.618	−0.3	0.382
0.4	0.655	−0.4	0.345
0.5	0.692	−0.5	0.309
0.6	0.726	−0.6	0.274
0.7	0.758	−0.7	0.242
0.8	0.788	−0.8	0.212
0.9	0.816	−0.9	0.184
1.0	0.841	−1.0	0.159
1.2	0.885	−1.2	0.115
1.4	0.919	−1.4	0.081
1.6	0.945	−1.6	0.055
1.8	0.964	−1.8	0.036
2.0	0.977	−2.0	0.023
2.2	0.986	−2.2	0.014
2.4	0.992	−2.4	0.008
2.6	0.995	−2.6	0.005
2.8	0.997	−2.8	0.003
3.0	0.999	−3.0	0.001

5. Select the probability for the Z factor from Table 11.4. (Table 11.4 is a range probability representative of a standard distribution curve.)

The summary of the calculation to establish the probability of reaching event 09 in 18.5 weeks instead of the scheduled 19 weeks is summarized in Table 11.3.

The actual probability obtained from Table 11.4, which is 0.242, means that the chances of reaching event 09 in 18.5 weeks instead of the 19.0 weeks is 0.242 out of 1.000. The same method of calculation can be made for any other schedule time that may be desired.

11.7 PERT Cost Control

The use of the PERT system to control costs of a program can be effective for whatever degree of detail desired. The most convenient breakdown of PERT costs is based on the material engineering, manufacturing, and other costs of the work packages. A reporting system which is correlated with the PERT schedule is used, whereby a continuing comparison can be made between incurred and estimated costs for each work package (activities and events).

It should be noted that the PERT cost-control system promotes efficient use of resources with the resultant economies. The resources scheduled for activities of slack paths can be temporarily reassigned, thereby avoiding nonproductivity while waiting for effort in the more critical paths to be completed.

Comparisons of the incurred versus the scheduled costs reveal to management the financial status of a project and permit timely action to review areas of cost coverages and to effect means to remedy causes of excessive costs.

11.8   PERT Reporting Documents

Various reports designed to apprise management of the status and trends of a project are prepared and distributed. A report such as the Management Summary Report depicts the cost status and projected time required for a project as of a specific date. Specifically, the report reveals the following types of information to management:

1. *Schedule Status:* The difference between the planned schedule and the expected schedule reveals the degree of slippage, if any.

2. *Trouble Areas:* The areas which pose a threat to either the planned cost or the schedule of a project are trouble areas.

3. *Cost Status:* The comparison of the actual costs against the budgeted cost reveals the cost status.

4. *Cost Prediction:* The extrapolation of the actual costs reveals whether a cost overrun or underrun will be realized.

The report provides information so that management can pinpoint the trouble areas and apply the necessary remedial effort in time to contain, minimize, or eliminate the causes for the expected difficulty to the project.

The personnel loading reports and displays are useful for planning the personnel requirements and resources for a project. These reports indicate weekly personnel requirements and permit management to meet heavily loaded periods by shifting and hiring personnel. Figure 11.4 presents a typical manpower loading display used for PERT systems.

Two other reports which are generally used for the PERT plan are the Cost Prediction Report (Figure 11.5) and the Schedule Prediction Report (Figure 11.6). Because of the close correlation between these two reports, they will be treated simultaneously. On these report forms, the cost and schedule information is plotted for each period in such a manner as to facilitate a projection of the trend of the cost and scheduling of a project.

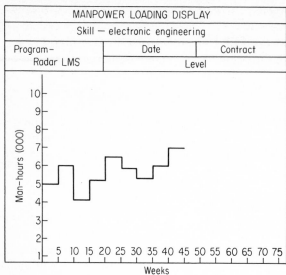

FIG. 11.4   Manpower loading display.

Figures 11.5 and 11.6 illustrate the Cost and Scheduling Reports for the Shadow Generating System of the RLS.

To illustrate the type of information derived from these curves, a hypothetical situation is assumed in which work on the project has proceeded to a particular point. The normal is represented by the zero axis,

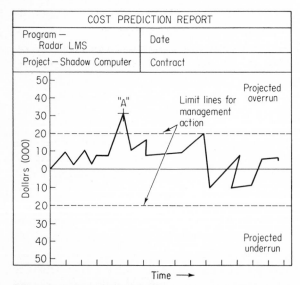

FIG. 11.5   Cost Prediction Report.

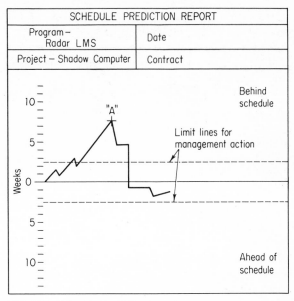

FIG. 11.6   Schedule Prediction Report.

and both curves should center around the axis. The dotted lines represent limits or tolerances, and if either curve penetrates the limit line (especially the line representing cost overrun or behind schedule progress), management action is mandatory.

The curves will usually indicate trends so that corrective action can be taken before the curve penetrates the limit lines. At point A, the curves indicate that an adverse situation was experienced and that corrective action was taken. Because of the corrrective action, the trend of the curves was reversed, and the project was brought back within the cost and schedule limits.

11.9   Line of Balance (LOB)

The LOB is a management tool akin to PERT except that it presents only the current project status and has no predictive features such as in PERT. When applied to a project involving a single unit or a small quantity, the LOB represents a status report, depicting the degree to which each of the disciplines associated with a project meet the schedule objectives that were established. Originally, the LOB technique was designed for and used on production projects involving large numbers of units. The concept was later modified and variations of the technique are now used for projects involving the design and development of single or small quantities of units such as with the RLS. Since this text discusses the role of the project engineer in managing projects involving engineering, the discussion of the LOB technique will center on principles relating to projects typified by the landmass simulator.

The LOB report for any point in time consists of the following graphic elements:

1. *Objective Chart:* A plot of the schedule for percent completed effort (or items in the case of production projects) against time

2. *Progress Chart:* A representation of percent completed effort (or items) against type of effort (or production plan for production projects)

3. *Plan:* A plot of types of effort (or production items) against time schedule

To illustrate how a LOB report is compiled and used, Figure 11.7 is presented for the Shadow Generator of the landmass simulator. Normally the LOB report would address the total Radar Landmass device, but in order to correlate the information with what was previously pre-

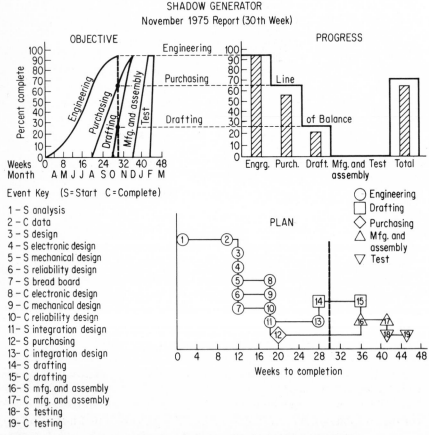

FIG. 11.7   Line of Balance for Shadow Computer of Radar Landmass Simulator.

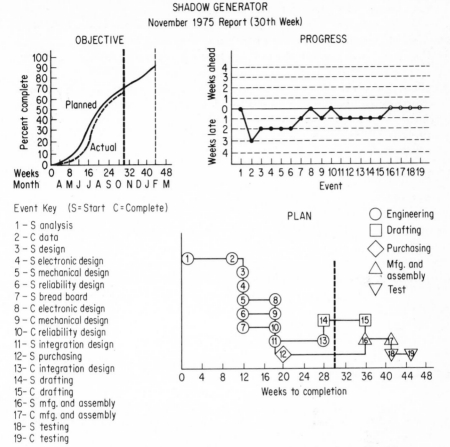

FIG. 11.8 Modified Line of Balance for Shadow Generator of Radar Landmass Simulator.

sented in discussing PERT and to avoid a complex presentation, only the above noted subsystem is discussed.

In the Objective Chart, the schedule curves for engineering, purchasing, drafting, manufacturing, and assembly and testing are shown. For the November report, the vertical 30-week line is shown and the status line representing the percent that should be completed for each type of effort at the 30-week point is carried over to the Progress Chart. The LOB for each type of effort is drawn as shown. The shaded vertical columns represent percent actual completion status. The total column represents the total effort accomplished for the complete Shadow Generator as compared to the LOB line for the system.

The Plan Chart presents the schedule for various events that are coded for the five categories of effort previously discussed. The identity of the various events was derived from the PERT chart, Figure 11.2, to permit the reader to correlate the principle discussed between the two reporting systems. Normally, if the LOB system is used on a project there would be no PERT system designed and the identity of the events as shown on the Plan Chart would be closer as judged necessary by the project engineer.

Figure 11.8 is another variation of the LOB reporting system as is often used for development programs. The Objective Chart depicts total effort scheduled and completed for the Shadow Generator subsystem as of the 30th week. The Progress Chart shows the status of accomplished effort for each of the events that are being tracked.

11.10   Summary

The PERT system is a mangement tool designed to control the schedule, cost, and technical performance variables of complex programs and is used primarily where devlopment effort is required. In the traditional organizational complexes, reports to management were historical in nature and therefore precluded timely corrective action on current programs. The PERT system offers a means of comparing current performance and status against planned performance, thereby revealing areas of difficulty and permitting timely corrective action on the causes rather than the symptoms of the problems.

The PERT cycle consists of the elements of determination of objectives, creation of plans, establishment of schedules, evaluation of performance and making of decisions. The decision-making effort involves a corrective feedback cycle which results in a very flexible medium to handle unanticipated developments and changes.

The work package is the basic building block of the PERT program and consists of a family of activities and events. The degree of control required will dictate the amount of detail and the size and number of the work packages.

The PERT network is a flow diagram which gives a graphic representation of the requirements and relationships for various disciplines from the point of view of time and effort and consists of activities and events. In an estimation of the length of time to complete an activity, the three figures of optimistic, pessimistic, and most likely figures are used.

The critical path of a PERT network reflects the path that requires the longest period of time to traverse. A slack path is one in which available time greater than the time required to complete the critical path is present.

The PERT plan is based on the statistical probability that the chance of meeting the schedule of earliest completion date for the program itself or any event is 50–50. The probability of meeting a date of an event other than the earliest completion date can be calculated by applying statistical formulas and using appropriate tables.

The PERT cost control is directly related to the time of activities required for work packages. Control of cost is accomplished in a manner similar to schedule control.

The PERT plan incorporates numerous reporting documents which serve to keep management informed of the program status and thereby to facilitate timely decisions for corrective action.

A LOB reporting system is appropriate for the smaller and less complex projects. It does not provide any predictive features but provides a useful management tool for providing the status of a project.

There are generally three elements that comprise a LOB report: the Objective, Progress, and Plan Charts which give management the essential information relating to the project.

PROBLEMS

The construction of a modern frame house would involve the tier II activities indicated below. The estimated length of time for each activity is indicated. An architect–builder has accepted the house project which must be completed in its entirety in 40 weeks from the time of establishment of the plan and design.

1. Prepare a PERT network for the project, indicating activities and events.

2. Calculate and indicate the optimistic, pessimistic, and most likely times for each activity.

3. Calculate and indicate the most critical path of the network.

4. Calculate and indicate the slack paths.

5. If it is evident that the 40-week time limitation will be exceeded, explain which activities can be accelerated, and explain the logic followed.

6. If unusual weather conditions extended the excavation time by 2 weeks and the time to build the foundation by 2 weeks, how would such developments affect the most critical path?

Activity	Weeks
1. Planning and design	12
2. Purchase of materials	7
3. Financing arrangements	4
4. Excavation	2
5. Foundation	3
6. Framework	13
7. Plumbing	4
8. Electrical work	5
9. Roof and enclosure	12
10. Electrical appliances	3
11. Plumbing fixtures	2
12. Interior work	6
13. Painting	7
14. Landscaping	4

7. Explain how revision of other activities might be effected to make up for the lost time described in question 6.

8. Calculate the probability of completing the project in 42 weeks.

Chapter Twelve

Initiating the Project

12.1 Review of Contractual Obligations

The signing of a contract that was received officially executes a legally binding agreement between the buyer and the seller. Project engineers must then initiate a contract-monitoring role which will continue until all items are delivered as required by the contract terms.

The framework of their organization, the scheduling, and the lines of authority have already been established by virtue of their work in preparing the technical proposal and deriving the bid cost figures. The primary and in some respects the sole objective of the project engineer is to deliver the items which meet the technical requirements within the cost and time schedule of the contract.

Because of the dialogue that had taken place during negotiations, the technical requirements for the equipment and the cost and the delivery schedule may differ in a number of areas from what was described in the technical proposal. The joint task that confronts the project engineer is to review the final documents that comprise the signed contract and realign the project plans, resources, and schedule to enable the company to perform in accordance with the official contract.

A specification, particularly if it is a performance type, may have had areas edited during technical clarification and negotiation. As a result, adjustments to the proposed price were probably made and revisions to the technical proposal agreed upon. Whereas the revisions are reflected in the contract, the project engineer probably may not have the time to revise the internal technical design approach resources to satisfy the areas that were clarified and revised. As a result, one of the first tasks that the project engineer must perform is to realign the technical approach to be responsive to the contract that was signed.

If for the RLS the negotiation clarifications brought out the fact that the side lobes of the antenna pattern were to be simulated, the project engineer would be required to provide for the necessary engineering effort and seek out adequate resources. The added complexity of that subsystem design might require a specialist and might entail some development work. The project engineer may thus have to reassign an engineer possessing the special experience and knowledge required for what in effect is a new technical requirement for the equipment. The same review of the schedule, facilities, and cost budget would be necessary to assure that the effort that is to be applied on the project will result in reducing equipment that meets the contract requirements.

The same review and adjustment would be necessary for the delivery schedule and money resources. For instance, the revised design approach may require time and effort which differs from the original proposal. The budget for man-hours and funds would have to be revised to permit satisfying the contract as signed.

The above emphasizes the importance for project engineers to review the final contract details to avoid the necessity of correcting errors at a later point in the project. They must also keep in mind that the order of precedence of contractual documents determines any point of conflict. Since the contract schedule and specification take precedence over the technical proposal, the project engineer has to recognize that the specifics expressed in the proposal at the time of solicitation would become contractually binding only in situations in which the contract schedule, specification, and referenced documents were silent on any controversial point expressed in the proposal.

12.2 Project Organization

Up to the point of contract award, the project engineer has, in essence, been performing as a staff engineer with only temporary delegated authority over other personnel of Acmen Electronics. The cost and time consumed in the proposal preparation, negotiation, etc., have been charged as general overhead expenses.

Upon receipt of the contract, a project team will be officially established. The project engineer will be given the authority necessary to execute the responsibilities required of this position, and an organizational plan will be formally announced to all employees of the company. The project organizational plan will vary for different contracts but must be designed to handle the specific objectives and anticipated problems that may arise. The organization for the RLS program as derived by Acmen Electronics is shown in Figure 12.1.

As noted in the organization structure, the project engineer occupies the key position, vested with the responsibility and authority over all members of the project team with the exception of the contract administrator and the Configuration Management Office, discussed in Chapter 13. The reason for the exceptions is that both organization units have responsibilities that extend beyond the domain of any single project so that their responsibilities would be compromised if they were subject to the project engineer's complete authority. Also, they perform a function of checks and balances which are essential to any organization.

The contract administrator is, of course, concerned with contractual matters which become heavily involved with legal issues. Since the interests of the project are vital to the company's welfare, the contract administrator will nearly always be guided by inputs from the project engineer. In cases where unresolved disagreements occur, the issues would be referred to the higher echelons of the company for decisions.

The Configuration Management Office would exercise jurisdiction in matters which relate to changes, audits, and the establishment of baselines. The authority of the office does not in any way compromise the authority of the project engineer since the Configuration Management Office, to a large degree, functions to verify that the project engineer's responsibility concerning changes, audits, and baselines are being satisfactorily carried out.

The project engineer has direct line authority over the functions of the project purchasing agent, production coordinator, design engineer, and quality coordinator departments. The information flowing from these four departments and the decisions and direction given to them are essentially technical in nature.

The schedule and cost coordinator receives information relating to the schedule and cost of the project from the four departments shown in Figure 12.1, collates this data, and presents it in a form that is readily compared with the budgeted cost and schedule of the program to the project engineer.

Information originated by the contract administrator is submitted to the project engineer. If the information concerns the schedule or cost of the project it is processed and submitted to the schedule and cost ad-

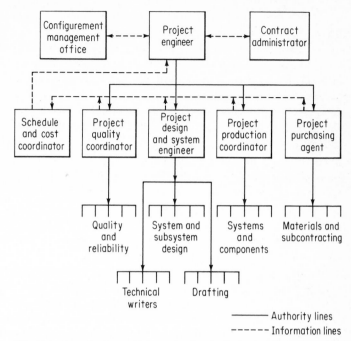

FIG. 12.1 **Organization and information flow for Radar Landmass Simulator project.**

ministrator for interpretation and integration in the status reports for use by the project engineer.

Each of the four departments, over which the project engineer has direct line authority, is responsible for the effort of branches or individuals assigned specific functions necessary to the program. The project design engineer, for example, is responsible for branches handling the system and subsystem design, the technical writing, and the drafting that are required in the program. This person must coordinate the work and the schedule for all branches in accordance with the broad parameters set forth by the project engineer.

12.3 Establishing of Task and Functions

Before making detailed assignments of work on the project, the project engineer will have already completed the updating of the equipment requirements, established the existence of resources for the job, and identified the organization of the project.

The identification of tasks usually uses some version of Work Breakdown Structure which the project engineer will recognize has been used in much of the preliminary work on the project. The design detail of Figure 3.1, for instance, represents a Work Breakdown Structure of the RLS. It was then used as the basis for the cost estimate that was discussed in Chapter 6. Elements of the Work Breakdown Structure were again utilized in working out the reporting systems for PERT and to a lesser degree, LOB.

In establishing the tasks for elements of the project, as identified in the Work Breakdown Structure, the project engineer will also establish the schedule for each task, the budget of man-hours and material costs, the performance criteria, and all other pertinent information that is necessary. In this respect, the information received from the various sources which were used in writing the proposal and preparing the cost estimate would be used to a great degree.

Since the landmass simulator is not the only project being processed by Acmen Electronics Company, the different assignments must be coordinated with the Master Corporation Schedule in order not to experience conflicts with other projects. In Chapter 4, however, the overall capacity of Acmen was presented as a part of the proposal, and the analysis indicated that there was ample capacity for the RLS program. With this in mind, it will be assumed that the complications presented by conflicts with other projects do not exist. In passing, it should be stated that conflicts of demand for the services of a specialized discipline is a very real problem in industry and has been responsible for large numbers of delays and difficulties that have been experienced on different programs. These situations are particularly prevalent when there is poor overall organizational planning or when slippage occurs on one program which results in jamming together with a different program that is scheduled behind the first one.

12.4   Scheduling of Project Tasks

The assignment of tasks, budgeting of expenditures of time, and establishment of goals require a high degree of coordination and scheduling among the various disciplines of effort. Ideally, the project engineer strives for perfect coordination where no person would remain idle while waiting for a particular task to become available or where no person, because of upset scheduling, has more work at a particular time than can be handled. In planning the scheduling of the tasks, the project engineer would prepare a project phasing chart similar to the one in Figure 12.2.

Figure 12.3 shows a phasing chart for the engineering and fabrication effort required for the simulators based on the tier III breakdown of Figure 3.1. The sequence of work follows an orderly and logical pattern. For instance, the drafting effort cannot be initiated prior to the comple-

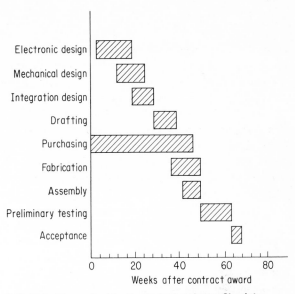

FIG. 12.2   Phase chart for the Radar Landmass Simulator.

tion of the preliminary engineering and design engineering effort. Also, there is a sequence of effort among the various systems. The phasing charts of the types shown are used by the project engineer in setting up the overall schedule of effort and are used as management tools for graphically portraying the overall schedule of the project or its divisions.

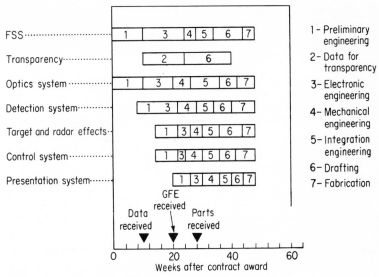

FIG. 12.3   Phasing chart for engineering and fabrication effort.

12.5 Make or Buy Decisions

During the precontract phase, the different factors related to whether a component, subassembly, or subsystem should be subcontracted were recognized, and the technical proposal indicated the intent of the offeror at the time of the negotiations (if the TPR called for such information). Unless the ultimate contract specifies that a subcontract be executed with a particular company for a particular item, the contractor has no legal obligation to enter into a subcontract agreement if events subsequent to the submission of the proposal indicate that the design and fabrication of a particular item could, in fact, be accomplished in-house.

In initiating the project, the project engineer must make the final decision as to what items, subassemblies, systems, etc., should be subcontracted. The various factors which previously dictated the decision to subcontract should be reanalyzed, and other questionable areas should be considered. This action is particularly necessary if, as often happens, there have been revisions in the procurement requirements brought about by the negotiations.

Some of the prime factors which influence the decision regarding subcontracting of a particular article are summarized as follows: contractor's capability, experience, relative cost, schedule of contract, loading, future activity, and customer's desires.

The capability relates to the engineering and manufacturing proficiency of a contractor in a particular area. In the case of the RLS, one area which might be considered for subcontracting is the optical system. The project engineer should recognize several fundamental problems related to the optical system that must be resolved. The first point is that optics systems must be designed in such a manner as to permit sufficient light to be transmitted from the FSS source through the various lens systems and on to the transparency itself. A problem associated with this light-transmission problem is that the design must be such as to keep the noise level to a minimum. Noise in this case is optical noise, or spurious light caused by dispersion of the FSS light as it travels through the lenses. Another critical point is that the spot of light impinging on the transparency must be very sharply focused to a dimension of 0.001 inch.

The project engineer must make an objective analysis of the capability and experience of the optical engineering department to determine whether Acmen Electronics Company is technically qualified to design the optics system. In like manner, the fabrication and assembly of the optical system, for which highly skilled workmanship is involved, must be analyzed. As a result of the analysis, the project engineer may conclude that the engineering design of the optics system by Acmen personnel is warranted, but that the fabrication and assembly of a system by a subcontracting firm would be the best course of action.

Cost is always a prime factor in any decision regarding subcontracting in any area. Everything else being more or less equal, it would be advantageous to perform the work in-house. However, because of experience, know-how, facilities, and other factors, one company may be able to design and/or produce a component or system at less cost than another company.

To demonstrate how the factors of cost would be analyzed by the project engineer in making a determination regarding subcontracting, a hypothetical case will be illustrated. If, for example, the contractor has determined that the direct costs and overhead for a particular task are $115,000, the problem is to establish the point at which it would be advantageous to subcontract. Assuming that the best offer from a subcontractor for the task is $100,000 for direct and overhead costs, the following tabulated analysis of the cost to Acmen is presented:

	Acmen produced	Subcontractor produced
Direct and overhead cost	$115,000	$100,000
G&A, 10%	11,500	10,000
	$126,500	$110,000
Profit, 10%	12,650	11,000*
	$139,150	$121,000†
Acmen G&A, 10%		12,100
		$133,100
Acmen profit, 10%		13,310
		$146,410

* Subcontractor's profit.
† Selling price to Acmen.

From the above table, it is seen that from the cost point of view, it would be less costly to perform the task in-house.

Another matter that must be considered before reaching a decision regarding subcontracting is the schedule of the program. A comparison between the times required for the prime contractor as opposed to the subcontractor to accomplish a particular task should be made in the same manner as the comparison of costs was made. As was noted in the analysis of costs, increments must be added to the subcontracted effort in order to arrive at a realistic comparison. The schedule increments are time for specification preparation, time for bid responses and subcontract award, time for acceptance testing, and liaison time.

All the increments of time as noted above must be added to the time that the subcontractor requires to design and produce a particular item or system. If, for example, the total of the additional elements of time noted above amounted to 8 weeks and the subcontractor's time required to produce an article is 30 weeks, the time cycle for subcontracting is then 38 weeks. If time were the only factor and the prime contractor could produce an article in 35 weeks, then the obvious decision would be to design and produce the article or system in-house.

Quite often, the loading of the prime contractor is such that the personnel and/or facilities are not available to produce an article in a timely manner. If the prime contractor does not wish to invest in additional capacity or hire additional personnel, an article would be subcontracted.

An important factor to consider is what future market a particular system or article offers. In spite of undesirable costs, schedule, problems, etc., the prime contractor may desire to produce an article in-house and thereby derive the experience and know-how for future business.

The preference of the customer is another factor that must be considered. A customer may express a preference or even an insistence that a particular item be procured from a particular subcontractor. Whenever possible, the prime contractor should cater to the preference or even whim of the customer if the cost, delivery, or performance of the project is not jeopardized. Usually, the customer has a sound basis for desiring an article produced by a particular subcontractor, and the project would not be adversely affected if such an article were used. If the prime contractor finds that by dealing with a particular subcontractor of the customer's choice, the project may be adversely affected, the undesirable effects should be made known to the customer as soon as possible, and an adjustment should be made in the contract cost, schedule, or performance requirement.

12.6 Assigning of Tasks

The information in Section 12.4, which dealt with the scheduling of the various tasks of the project, is the basis for the assignment of tasks to the various departments in the project organization. A document identified as a Task Assignment Form would be prepared by the project engineer and issued to each of the four project departments and to the schedule and cost coordinator. There are many variations of the Task Assignment Form, both as to what it is called and as to the form for the required information. The form which is illustrated in Table 12.1 calls for basic information that is common to practically all such documents and in this particular case describes the task assigned to the project design engineer.

The Task Assignment Form designates what is required, when it is required, and what the budget to accomplish the task is. There are

TABLE 12.1 Sample Task Assignment Form for Project Design Engineer on Radar Landmass Simulator Project

---

ASSIGNEE: Project Design Engineer, K. Lawson
TYPE OF TASK: Design of Radar Landmass Simulator
DATE: 15 January 19-
PROJECT: Radar Landmass Simulator Device 5A1
QUANTITY: One (1) Prototype
PROJECT NO.: 6853 Contract No.- - -N652A
DESCRIPTION OF TASK: Design the Radar Landmass Simulator which conforms to customer's specification (Attachment 1). Design approach shall be consistent with that described in Proposed Task Design, Project 6853 (Attachment 2).
TASK SCHEDULE: The schedule of the task to be initiated at once is to be completed as required by Task Schedule shown in Attachment 3.
BUDGET: Engineering hours—as shown in Attachment 4.
REPORTS: Cost Prediction Report (Monthly)
     Line of Balance Report (Monthly)
     Design Status Report (Semi-Monthly)
     Manpower Loading Display (Monthly)
ATTACHMENT (1) Specification No. 1001
ATTACHMENT (2) Proposed Design Approach, Radar Landmass Simulator, Project 6853
ATTACHMENT (3) Task Schedule, Project 6853, Engineering Design: Radar Landmass Simulator
ATTACHMENT (4) Budget: Engineering Design, Radar Landmass Simulator, Project 6853

---

numerous supplementary documents which provide the necessary detail for the task and also indicate in many cases how the task is to be performed. In addition to the Task Assignment Form, there is a constant communication flow, both written and verbal, to clarify the task requirements.

The project design engineer will in turn divide the task into subtasks and make assignments to various subordinates for execution, giving the necessary information, direction, schedule, and budgets that were provided. The subdivided tasks would not only relate to the various areas of design such as identified in the tier II or III detail shown in Figure 3.1, but also would be divided into disciplines of engineering effort. For instance, the Shadow Computer would involve electronic design effort and mechanical engineering effort for meeting packaging and other requirements.

The Task Assignment Form for the project purchasing agent will contain a different type of task designation than that contained in the form for the project design engineer but will provide the basic information relating to the schedule, objectives, budget, and other requirements. Of particular interest to the project purchasing agent will be those areas of subcontract in which negotiations must take place as contrasted to the routine task of ordering standard parts such as transistors. The project purchasing agent would desire to initiate negotiation action for subcontractors as soon as possible and is therefore anxious to obtain information in such areas as soon as possible.

12.7 Summary

The project engineer who is responsible for the successful prosecution of a contract, has line authority over the different areas of required effort. Such authority is depicted in an organizational chart in which authority for the project emanates from the project engineer.

The schedule objectives and budgets for the various areas of effort are established by the project engineer. The overall schedule of effort on the project must be phased to avoid either the idleness of persons waiting for some prior work to be completed or a jam-up of work due to several tasks reaching a group at one time.

One particular phase of the project that relates to subcontracting is the weighing of the various factors involved to determine whether a task should be subcontracted or done in-house. The main considerations are the capability, cost, schedule, and capacity of the prime contractor.

Once the organization for the project and the schedule for the accomplishment of various tasks are established, the project engineer executes the Task Assignment Forms for various main departments such as project design engineering, purchasing, manufacturing, and quality control, which formalize what is to be accomplished, what the task completion date is, what the budget contains, and in many cases how the task is to be completed.

The initiation of the project therefore entails establishing the organization, schedule, budget, and assigning of the various tasks to the key organization personnel.

PROBLEMS

1. Draw up an organizational place for the Science Manufacturing Company to handle the gas ignition system described in problem 1 at the end of Chapter 6.

2. Set up a schedule of effort indicating areas of subcontract.

3. Set up task assignment sheets for the major area of effort for the project of the gas ignition system.

Chapter Thirteen

Configuration Management

13.1 Introduction

Whether they be kitchen appliances or sophisticated missile systems, the products of modern technology and engineering innovation are becoming increasingly automated and complex. During the time cycle of conceiving, defining, producing, and using a particular system the project engineer and other management officials will be faced with making decisions as to whether or not to adopt a change to the hardware design and if a change is adopted, how to handle the associated needs. Some of the more common types of proposed changes that require decisions include the following:

Change in performance capabilities
Change in appearance, size, weight, etc.
Change in design
Substitution of new modules or components
Change in computer program

In the pre-World War II era, the requirement to consider changes for less complex products occurred infrequently. When it did occur, a company was able to get away with an informal system for handling such revisions and modifications. However, for the more complex equipment of that era, the efficient and well-managed companies devised systems for controlling and documenting changes which embraced the underlying principles of Configuration Management as widely practiced today.

During the initial phases of the space program, the government authorities soon recognized that loosely controlled changes not only caused complications in their highly complex systems but raised havoc with other systems and equipments that were designed to operate in close coordination with each other. Also, since the consideration of changes often was required at any point in time during the life cycle of the equipment, it was recognized that the documents such as drawings, which identified the equipment at that point, had to completely and accurately describe the equipment. Many of the complex projects that were pursued in the space programs were experiencing cost and schedule difficulties because the traditional methods for handling changes were not working. As a result, the formal Configuration Management techniques that were introduced and used were in large measure responsible for the successful management of the numerous space projects that were pursued and ultimately completed.

The implementation of Configuration Management techniques was justified since it was recognized that changes of any type could disrupt the progress of a program, upset schedules, and add significantly to the costs—particularly if the change required redesign or refabrication. Because of the effects that changes might have on a project, the Configuration Management Plan requires that each proposed change be evaluated to establish whether the merits of the change are worth the cost and time that it entails. To assure the objective evaluation of proposed changes, the Configuration Management Plan requires that a Change Control Board be established. The Board reviews all proposed changes on a particular project, establishes how essential a change may be, evaluates the impact of the change on cost and schedule, and determines if the change is to be incorporated, postponed, or rejected. Once the Control Board decides that a change be adopted, the other disciplines of Configuration Management to be discussed come into play.

13.2 Configuration Management Principles

A broad definition of Configuration Management could be stated as follows:

> The implementation of formal management and technical direction and controls during a project's life cycle to provide a complete definition of the function and physical characteristics of each item, control the adoption of changes and maintain a continuous accounting of the design and equipment.

Under the Configuration Management Plan, the life cycle of a project is divided into the following four basic phases:

1. Concept formulation phase
2. Definition phase
3. Acquisition phase
   a. Design and development stage
   b. Production stage
4. Operational phase

As shown in Figure 13.1 (see p. 161), each phase terminates at the baselines identified as follows:

1. Characteristics baseline terminates the concept formulation phase.
2. Functional baseline terminates the definition phase.
3. Operational baseline terminates the acquisition phase.

In addition, the Product subbaseline terminates the design and development stage of the acquisition phase.

There is no identified baseline that terminates the operational phase other than to note that at the termination of the operational phase, the equipment is scrapped or reconfigured to the degree where it might be said that the reconfigured equipment assumes a new identity and requires a new Configuration Management Plan.

The baselines represent checkpoints during the life cycle of the project. The initiation of a new phase cannot be made until all details and questions that have been raised during the previous phase have been resolved. For example, if during the concept formulation phase the planners for the RLS were undecided as to whether the size of the simulated area was to be 1,500 by 800 miles square or 500 miles square, the next phase, which is the definition phase, could not be initiated since the characteristics baseline had not been established. However, when the point in question is resolved, the documents describing the characteristics baseline would specify the area to be the 1,500 by 800 mile area, and the definition phase can be initiated.

In addition to the Configuration Management facets of phases and baselines, the disciplines of identity control and accounting must be applied.

The identification discipline requires that during each phase, the complete and accurate description relating to the products of the particular phase be documented. For example, the products of the definition phase would be the specification, schedule, contract, etc., which comprise the functional baseline of Figure 13.1.

Control is the second discipline of Configuration Management. Proper control assures that whatever is done during a particular phase is approved by the authorized officials and is accurately and completely reflected in the baseline documents.

The objective of the accounting discipline is to establish an orderly system and procedure for documenting all descriptions and revisions so that complete and accurate data relating to the equipment can be retrieved when desired. Part of the accounting discipline is the audit procedure which is implemented to establish that the data describing the equipment is both accurate and complete.

13.3 When Configuration Management Is Required

Practically any company involved in producing equipment has in its system some form of Configuration Management—particularly if engineering and revisions are normally involved. The degree and scope to which it is applied is often inadequate. Some of the major symptoms of a project which would suggest that Configuration Management disciplines and procedures are not effectively applied would include the following:

1. *Delayed decisions on changes*: When the question regarding the merits of a recommended change is raised, a delay in making the decision has an adverse effect regardless of whether the decision is yes or no. Work will be progressing and if the ultimate decision is to approve the change, the effort and associated costs to implement the change are usually greater due to the decision delay. The formal procedures that are contained in a Configuration Management Plan promote prompt decisions regarding changes.

2. *Excessive cost of changes*: Changes are often approved without a comprehensive analysis of the cost impact of such changes. The procedures inherent in a Configuration Management Plan require an analysis of cost trade-offs. If the predicted cost of the change exceeds the value or advantages that would result from the proposed change, it would not be approved.

3. *Documentation not consistent with equipment*: The failure to make timely revisions to drawings and literature reflecting the changes that were made to the equipment or computer program usually results in confusion. Time-consuming analysis of the documentation that is not consistent with the equipment is required, particularly when the equipment requires servicing. The procedures provided by Configuration Management assure that the revised equipment be accurately and completely supported by revised documentation.

4. *Objectives not clearly stated:* Engineering effort, particularly when related to R&D, is not directed to the requirements of the project. Configuration Management defines the specific objectives at the start of a phase and minimizes the possibility of unnecessary effort which does not follow a direct path to the objectives that were established.

13.4   Concept Formulation Phase

As the name implies, this phase involves, to a large extent, abstract planning and translation of requirements into the equipment specification. In the case of the RLS, the concept formulation phase established that there was a need for some means to train radar operators how to operate and interpret the intelligence that the radar would provide in an actual mission. It can be assumed that the merits of different alternatives were evaluated such as using video tapes of actual missions, providing photographs to students, flying training missions, and other possible options. Out of all the alternatives considered, the RLS described in the specification that was discussed was selected.

Thus, it is during the concept formulation phase that the desired objectives for satisfying a requirement are considered and established. If the desired objectives are not feasible because of technical limitations, a research or development program may be pursued to provide a feasible technical approach. The concept formulation phase generally involves the consideration and rejection of ideas, probing, and creative thought. However, the Configuration Management Plan forces the desired objectives to be held in focus for all participants.

The culmination of the concept formulation effort is the establishment of the characteristics baseline which is represented by some form of requirements document. Until the baseline is reached, no type of effort relating to the subsequent definition phase is initiated. Either a Configuration Management administrator or committee would be charged with the responsibility for certifying that the documentation relating to a piece of equipment is acceptable and meets the criteria for establishing the characteristics baseline. It should be noted that the formulation and definition phases are implemented by the customer and the acquisition phase by the contractor. The same individual would not function as project engineer on all phases except when an organization undertakes all phases of a project as a total effort.

13.5   Definition Phase

The primary objective of the definition phase effort is to translate the requirement that the characteristics baseline establishes into the equipment specification and other planning documents that will satisfy the requirements for the functional baseline as conveyed in Figure 13.1. The type and detail of the documents created during the definition phase will depend upon the particular project that is involved. If the equipment will be procured from a contractor who performs the design, development, fabrication, testing, etc., the documents that represent the functional baseline will include such documents as a performance and contract end item specification, contract schedule, proposal requests, etc. If, however, the equipment will be produced as an in-house effort, the baseline will include a design specification, a detailed plan for executing the various operations of the in-house effort, etc.

Any piece of equipment or system that is to be created must be supported by items such as operation manuals, maintenance handbooks, drawings, and spare parts. The number of support items is usually proportional to the complexity of the equipment. A kitchen toaster, for instance, will require only a simple booklet of instructions, whereas a complex computer system must usually include drawings, training courses, special test equipment, etc. All of the items of a project must be defined and scheduled during the definition phase. In Configuration Management, such items are called Contract End Items (CEIs) and they are described in the Contract End Item Specification. For the RLS project, the CEIs would be designated as the actual line items of the contract as described in Schedule Section A, Section 5.1 of the text.

Some characteristics of end items are as follows:

1. Capable of being described by a specification
2. Identified by top drawings and other definable nomenclature
3. Capable of being documented
4. Capable of being correlated with other end items

Some of the most common types of End Item Specifications are prime equipment, facilities, and support.

13.6 Acquisition Phase

The word acquisition infers that the equipment will be procured under contract. Whereas contracting is the usual means for obtaining equipment, the disciplines involved for the acquisition phase apply also to an in-house effort for designing and producing an item. For purposes of this discussion, acquisition by contracting will be discussed.

As noted in Figure 13.1, the acquisition phase is divided by the product subbaseline into the design and development and the production stages. The various documents that represent the functional baseline would be used to launch the design and development stage. During this stage of the acquisition phase, the products of the engineering, breadboarding, drafting, and other similar types of effort would be documented as engineering and manufacturing drawings, production specifications, and production planning schedules; when approved, such data would constitute the product subbaseline. Once the product subbaseline criteria have been approved, the production stage can be initiated.

In most projects, establishment of the product subbaseline does not

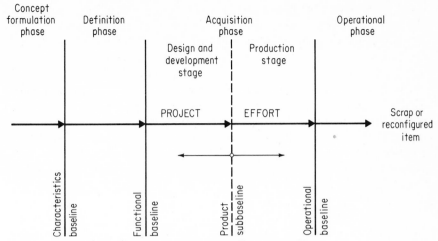

FIG. 13.1   Phases and baselines for Configuration Management Plan.

occur at a single point in time. The design of components and subsystems occurs at different rates and the documents and material required for the fabrication of one system may be ready at a different date. For instance, the detectors for the RLS may be designed and be ready for production well ahead of the optical system. The product subbaseline is usually staggered with respect to time. However, before any subsystem can be released for manufacture, the project engineer would have to verify that the requirements for the product subbaseline for the particular system in question is acceptable.

The primary effort during the production stage relates to the manufacture, fabrication assembly, testing and acceptance of the equipment required by the project. The operational baseline is established when the CEIs and, in particular, the equipment under procurement are delivered and accepted and when verification is made that the acquisition phase objectives that were documented at the functional baseline are satisfied.

13.7   Operational Phase

The utilization of the equipment that has been produced during the earlier phases occurs during the operational phase. The primary test as to whether the Configuration Management Plan was successful is whether the physical and performance characteristics of the equipment satisfy the objectives that were established and approved at the operational baseline. Another major objective of Configuration Management is to provide consistency between the equipment and the supporting documentation. If the drawings, manuals, and other support items completely and accurately reflect the equipment that is being utilized, then the objectives have been satisfactorily met.

The disciplines that would be applied during the operational phase would continue for the life of the equipment. For instance, changes to the equipment must be reflected in the drawings and other types of documentation. The Configuration Management effort would cease only when the equipment is scrapped or a redesign effort of such scope is applied that the equipment loses its former identity. Under the latter circumstances, an entirely new set of objectives would be established and a new Configuration Management Plan adopted.

13.8 Disciplines of Configuration Management

The three disciplines of Configuration Management, as cited in Section 13.2, are identification, control, and accounting. During the early phases, identification plays a major role since it is during the concept formulation and to a lesser degree during the definition phase that ideas, concepts, and objectives are formulated. The control and accounting disciplines can only be applied when the concepts, etc., take form and are documented. Therefore the control and accounting disciplines become dominant during the acquisition and operational phases.

Figure 13.2 presents a flow diagram of the identification process that might be used to establish technically feasible objectives during the concept formulation phase. The critical effort is the evaluation of technical

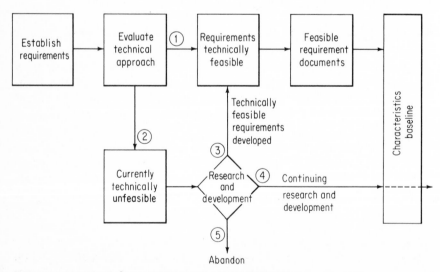

FIG. 13.2   Identification process during concept formulation phase.

approaches for a set of requirements. It is at this point that one of the following five alternatives can ultimately be taken. The alternatives are keyed by numbers to the different paths depicted in Figure 13.2 as follows:

Path 1: All the capabilities or requirements are technically feasible and can be documented to establish the characteristics baseline.

Path 2: The existing technology is considered inadequate for meeting the requirements that were established and an R&D effort will be necessary.

Path 3: The results of the R&D effort were successful in achieving a technology that was capable of satisfying the requirements, and therefore the project can continue toward the original requirement objectives.

Path 4: The R&D effort did not result in deriving a feasible technical approach for one or more of the required characteristics. In such a case, the project can continue on the basis that there exists a high level of confidence that technically feasible approaches for the requirement can be successsfully developed during the definition phase. However, the cost and risk of successfully developing the technical solution for the requirement that was judged unfeasible due to technological limitations increase as the project progresses into the definition phase. An organization must weigh the possibility of expending time and money for a technical solution that may not be achieved against the potential profits that would be realized if the R&D effort proves successful. Failure to achieve a breakthrough would probably mean that the project would have to be aborted.

Path 5: The fifth possibility is the judgment that some of the requirements are not feasible and cannot be developed within any realistic time frame and the requirement in question must be abandoned.

The point to be stressed is that by using the characteristics baseline as a checkpoint, the possibility of spending time and money on a project, expecting results that may not be technically feasible, is all but eliminated. Also, the area and objectives of R&D effort are identified and a time frame for achieving such objectives are established.

During the definition phase, the identification discipline becomes more concrete. Identification comprises the translation of the documented requirements that have been established as technically feasible into specifications and other planning documents.

For the acquisition phase, identification involves the translation of specification and contract documents into engineering drawings, job orders, and ultimately into the equipment itself. For the operational phase, the identification discipline is minimal since it would be considered primarily with changes to the equipment which must be reflected in the drawings and other documentation.

The control discipline in Configuration Management relates primarily to the direction of the effort necessary for achieving each baseline and evaluating, selecting, and monitoring changes that may be implemented.

The accounting discipline is closely related to control and is most heavily used in the acquisition and operational phases. The accounting disciplines, often implemented by a computer, serve to provide the following:

1. Maintain documents up to date by recording changes as they occur

2. Provide a current record of data in order to provide capability to retrieve information as required

3. Implement configuration audits to verify that the documents which describe the equipment are complete and up to date

As noted above, one of the accounting facets is the configuration audit that is conducted to certify that the configuration and performance of the equipment with its subsystems are consistent with the baseline documentation. Audits are normally conducted at the various baselines as a procedure for approving the baseline. Minor audits relate primarily to end items and can be conducted at any point in the life cycle of the project. Since audits generally involve hardware or other tangible items, they are more commonly conducted at the product and operational baselines.

The product baseline audit that occurs during the acquisition phase serves to achieve the following:

1. Compatibility of the functional and physical characteristics of the equipment or end items with the documents

2. Determination of the validity of the testing of the equipment serving to verify the specified requirements

3. Determination as to whether the documents such as drawings, etc., reflect all approved revisions and changes

The operational support baseline audits are conducted to achieve the following:

1. Establish that the drawings and data completely and accurately reflect the equipment including approved changes

2. Identify areas of change and revisions that are required

The responsibility and authority to conduct audits rest with the project engineer. On projects which include Configuration Management personnel, the project engineer would delegate the audits to such an official.

13.9 Changes

One of the major objectives of Configuration Management is to assure that when changes are made to the equipment, such changes are accurately reflected in the documents that support that equipment. The chaos and frustration that results from the situation in which the drawings or other documents do not accurately describe the equipment can be a major cause of unnecessary, time-consuming expenditures.

Figure 13.3 illustrates the basic procedures that are involved in assuring that approved changes to the equipment are faithfully reflected in the documents that support the equipment. As noted earlier, the characteristics baseline establishes the requirements for the equipment. It is the objective of the Configuration Management Program to assure that if any approved changes are implemented during the life cycle of the equipment, the documents of the characteristics baseline be updated as appropriate to reflect such changes. If, in the case of the RLS, a change was approved during the definition phase which increased the area to be simulated, such a change would be fed back to be reflected in the characteristics documents. The procedure thus assures that there will be no inconsistency between the documents that will eventually serve for the functional baseline.

The same procedure would be followed during any of the subsequent phases or stages. In summary, the results of the procedures for handling changes during the Configuration Management life cycle include the following:

1. During any phase, consistency among the documents and/or equipment being created during that phase and the previous baseline

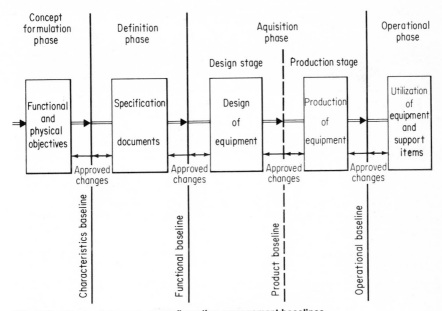

FIG. 13.3 Effects of changes on configuration management baselines.

2. Consistency among the documents that serve as the various baselines

3. Consistency between the equipment and support items that are utilized during the operational phase and the documents that serve as the operational baseline

The primary justification for implementing Configuration Management for a project of an organization is that overall costs will be reduced. The disciplines of identification, control, and accounting can be tailored to the size of the project. Configuration Management can be effectively applied to all sizes of organizations or projects. Some of the benefits that can be realized are:

1. *Establishment of objectives for each phase:* The baselines and the disciplines that are applied to such baselines eliminate the expenditure of effort on incorrect approaches or for unfeasible objectives. Once the baseline has been formally approved, the objectives for the next phase have been documented and the possibility of proceeding down the wrong path has been all but eliminated. The establishment of each of the successive baselines points the way to the next set of objectives.

2. *Effective channeling of resources:* The benefits derived from the channeling of resources are particularly valuable during the concept formulation phase since the effort relates to abstract notions and requirements. In the other phases, the channeling of resources is facilitated since the establishment of each of the baselines points the way to the objectives of the project.

3. *Elimination of redundant effort:* The clear establishment of objectives and channeling of resources act to eliminate redundant effort. Redundant effort usually involves two forms: work that more than one group is pursuing and work that must be repeated because of an erroneous approach or an erroneous objective. The implementation of the various controls such as design reviews and audits acts to detect errors and wrong approaches before they get too far.

4. *Facilitate retrieval of accurate data:* The Configuration Management change procedures require that only approved changes be made to the equipment and when they are made, the documents be revised at the time of change. The accounting discipline which provides for the retrieval of equipment data when required serves to eliminate wasted effort and unnecesary expense.

13.10   Summary

Configuration Management represents a formal way to track and record approved changes to the equipment being designed and built, and serves to channel resources to the objectives that have been established. There are four phases to a Configuration Management Plan, namely, (1) concept formulation, (2) definition, (3) acquisition, and (4) operational. In addition, the acquisition phase is divided into two stages which are the design and development stage and the production stage.

Baselines, which separate the different phases and stages, represent points in the Configuration Management cycle which provide a precise identity of the results of the earlier phase and the objectives of the following phase. The baselines are identified as the characteristics, functional, product, and operational baselines. The effort on a project cannot proceed until the documents constituting each baseline are verified as complete and accurate.

The three disciplines that are applied to varying degrees in each phase and stage are (1) identification, (2) control, and (3) accounting. A computer to peform the control and accounting functions is generally used on larger programs.

Chapter Fourteen

Project Monitoring, Communication, and Decision Making

14.1 Communicating

In order to successfully prosecute and complete a project, it is mandatory that an effective system for communication among assigned and interested people be provided. Three of the most vital reasons for an effective communication system are to provide information for decisions, to issue instructions or guidance, and to provide a means to report on the project status.

The communications for soliciting a decision usually are initiated by the existence of a problem that must be resolved. Figure 14.1 illustrates the dual communication cycles in the management and working-level areas and the various elements that comprise each cycle which relate to a typical project. It should be noted that the one element which is common to both cycles represents the project engineer's instruction function. The project engineer thus represents the key link between the two levels of the program.

Figure 14.1 shows only the internal communication cycles relating to response to higher authority or to the issue of directives. The project engineer will also be required to communicate beyond the organizational structure of the project both within and beyond the company. Much of this external communication would involve consultation services, specialist aid, and similar inputs. For instance, a special contractual problem may develop on the project. In order to pursue its resolution, the project engineer may require legal help which would be available from the corporate legal staff. The bulk of the external vital communication traffic is with the customer and involves discussion and correspondence relating to a multitude of points of a technical, contractual, and general nature. The smooth conduct of a project will depend to a large extent on how effective the communication link between the customer and the project engineer is monitored.

The project communication cycle of the working level would occur in the following sequence:

1. The instruction element is initiated by the project engineer.

2. The party responsible for putting the instruction into effect is represented by the application element. If the carrying out of the instruction does not meet with any problem, the action cycle is concluded.

3. If, however, as is frequently the case, the carrying out of the instruction meets with problems, the problem is identified as indicated in the cycle element titled problem and further cycle action is necessary.

4. The problem which is identified must be analyzed as indicated in the analysis element. The analysis element involves the gathering together of information relative to the problem which might require inputs from other sources not shown on the illustrated communication cycle.

5. After the analysis is completed, the evidence is studied and the decision is made.

6. The decision is then normally translated into an instruction, and the cycle is repeated. If everything worked well, the application of the instruction would terminate the communication cycle in question.

Occasionally project engineers may not be able to translate the decision into an instruction because of complications contributed by factors over which they have no control. For instance, the decision might be to utilize certain facilities of the company at a particular time. It may be that a conflict for the facilities exists with another project in the company. Therefore, the project engineer would refer the problem to the management-level cycle shown in Figure 14.1. The decision rendered in the management-level cycle would be referred to the project engineer for converting into an instruction for the working-level cycle.

In addition to communicating down through the project organization or up into the management domain, the project engineer must interface with organizations external to the company or with other company groups not associated with the project. The necessity for communicating with the customer, consultants, or potential subcontractors represents examples of external dialogues that might take place. Also, the project

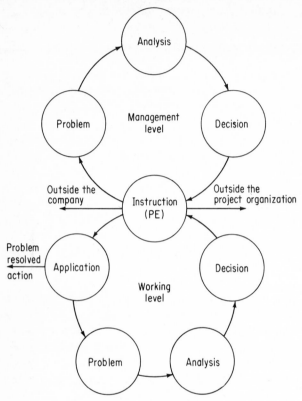

FIG. 14.1 Internal communication cycles for project monitoring.

engineer quite often must communicate with other departments such as accounting, personnel, etc. Figure 14.1 depicts the external communication paths.

It should be mentioned that many of the communication elements of the working level would probably be performed by one person. For instance, the evaluation and decision as well as the instruction elements would be performed in all likelihood by the project engineer.

14.2 Program Status Communication

Communications to reveal the status of a program flow from the working level up to the top echelon of management. The four basic points of information that the status reports must treat are:

1. Are the technical requirements of the specification being met?

2. Are the expended costs within the budget estimate or contract costs?

3. Are the schedules of the program being met?

4. Is there any indication that the requirements, schedule, and cost goals will not be met?

The status reports to the lower levels of management would involve a significant amount of detail since most of the corrective action, in the event of any project difficulties, would have to be taken by the lower management levels. At progressively higher levels of management, progressively less detail would be required.

Some companies conduct periodic project review conferences during which the project engineer personally presents the status of the project to assembled management officials. However, these oral presentations require the use of charts and other visual aids which constitute excerpts from the formal status report documents that are used in the organization.

The details of the forms vary among companies, but the forms provide for reporting the same fundamental information. In Chapter 11 the tools for management reporting were discussed. Whatever reporting system is being used on a particular program, the same general reporting means can be implemented. One fundamental type of report which is essential for providing the required status information is identified as the Management Summary Report discussed in Chapter 11. Other types of reports used to furnish different types of information are the Manpower Loading Display (Figure 11.4), the Cost Prediction Report (Figure 11.5), the Schedule Prediction Report (Figure 11.6), and the Line of Balance (Figures 11.7 and 11.8).

There are many other variations or types of reports that can be implemented, and it would be up to the project engineer to design and adapt any particular type deemed essential to the project. One thing that should be avoided is the temptation to adapt reports for a multitude of conditions, thereby building up a "paper mill" which will require so much effort for completing reports that the program will suffer.

14.3 Instructions and Direction

The communication of instructions by their nature extends from the higher levels of authority downward. Most instructions, especially at the higher level of management, are issued verbally. There have been volumes of material written on the various aspects of instruction communication, and it is not the purpose of this text to discuss the philosophy or techniques of verbal communication, except to state that any form of communication must be sufficiently precise, complete, and lucid to guarantee that the maximum possible degree of understanding of what is being communicated exists among the parties involved.

Instructions relating to tasks, policies, and other formal matters must be documented. Even verbal instructions, mentioned above, should be confirmed in writing if the instruction relates to any matter that is significant. Whereas general instructions would take a normal narrative form, instructions of a technical nature would be issued on a standard format. The use of a format guarantees that the basic information is included.

The authority that is extended to the project engineer to issue instructions and directions carries with it the responsibility for the results of such directions. The supervisor–subordinate relationship that is essential for effective management is a subject of its own, but it should be noted that the project engineer is given a function requiring a team effort of all individuals assigned to the project. Project engineers must therefore exercise the fundamental principles of human relationships with their subordinates.

In giving direction to the project team, the project engineer must make sure that the following major technicalities are satisfied:

1. The direction will not lead to a deliverable item which does not comply with the contract specification.

2. The direction represents the most realistic and reasonable cost to the project budget.

3. The direction can be accomplished with existing or available personnel and resources.

4. The direction will not require that the planning and delivery schedule be compromised.

The directions that the project engineer may give to parties external to the project organization or company have even more serious implications if the directions are faulty. Internally, actions to compensate for erroneous direction can be taken. The outside world, however, often will not be diligent in compensating for erroneous directions and may even take advantage of such errors to support a legal claim. For government and many types of commercial contracts, the principles discussed in Chapter 10 on Constructive Change Orders demonstrate the extent of diligence that project engineers should exercise in giving direction to the outside world on their project.

14.4 Engineering Instructions

One type of engineering instruction was discussed in Chapter 12 with regard to the Task Assignment Form. Project design engineers will carry out the directive as expressed in the Task Assignment Form by dividing the total engineering task into areas of effort to be subdivided among the various design groups under their direct supervision. To each of the design groups, project design engineers will issue an EO (Engineering Order), which sets forth the same basic information that is contained in the Task Assignment Form but describes the particular function in greater detail.

The issuance of the EO does not constitute an end in itself as far as the particular directive is concerned. Although the EO attempts to convey all the required information in the clearest and most complete manner that is possible, questions of interpretation and explanation always evolve. Thus, the project design engineer and the various group leaders to whom the EO is directed will invariably engage in many discussions on the assignment. If a problem arises that cannot be resolved by the project design engineer, it would be referred to the project engineer and the working-level communication cycle illustrated in Figure 14.1 would be traversed.

During the course of a major project, changes are almost inevitable. A change may be initiated internally or may be initiated by the customer. At any rate, any change must be reflected as a revision to the design, and appropriate information must be transmitted to the design engineers. The means used to incorporate a change in design is the EC (Engineering Change Form). The EC would provide all the technical, cost, and schedule information that is desired for implementation of the revision to the product in question.

14.5 Quality-control Instructions

In the project organizational chart shown in Figure 12.1, it is noted that quality-control functions are divorced from the manufacturing, even though quality control is intimately related to manufacturing. Aside from the fact that quality-control disciplines are specialty fields, the functions are separated as noted to preclude any conflict of interest since, if properly pursued, they would impose a stringent policing action on their related areas of endeavor in manufacturing.

Quality control is a complex and special function of any organization and is pursued in many different ways. The assembly-line operation would adapt a quality-control program significantly different from a project involving the development of a single or small number of items such as the RLS. However, certain fundamental decisions must be made by management and passed down as instructions for the quality-control department to carry out.

For the RLS the project engineer is primarily concerned with meeting the accuracy and performance requirements of the specification and with monitoring the critical systems of the simulator to guarantee the achievement of the requirements. An example of one accuracy requirement is that the simulated radar presentation be accurate to within ±3 percent in range. What is meant here is that the simulated radar range reading between the position of the aircraft and some specific point would be compared to some known standard. In the case of the dual transparency simulator, the standard would be a calibrated range scale on the transparency.

The project engineer in consultation with the project design engineer, the systems engineer, the project production manager, and the project quality coordinator would determine what systems and elements of the simulator must be controlled and what accuracies must be obtained in each system or element in order to achieve the ±3 percent range accuracy. The results of the team analysis might indicate that the design of the following systems must be considered for achieving the specified range accuracies:

1. *Flying Spot Scanner:* Electronic circuit by which the accuracy control of the FSS spot is achieved
2. *Transparency:* Accuracy with which the geographic information is stored on the transparencies
3. *Control of radar display:* Accuracy with which the simulated radar display is read using the radar control circuit

Having established the information that is to be transmitted, the project engineer would document the pertinent instructions in the Quality Control Order (QCO) to the design, production, and quality-control departments for execution. The project engineer will look to the project quality coordinator to guarantee that the simulator design make possible the achievement of the ±3-percent range accuracy and that the fabrication and assembly of the unit be accomplished in such a manner that the design objectives are realized. The accomplishment of the objective involves communication of the proper instructions to the responsible parties, a follow-up system of developments, and communication of corrective instructions and information among all parties.

14.6   Drafting Instructions

There are two basic schools of thought about the organizational status of the drafting effort. One holds that since drafting is so closely related to engineering, the effort should be integrated with the engineering department itself. The other is that all drafting should be accomplished in a separate organization. There are many arguments that can be advanced for each concept. Most companies, especially the larger organizations, have adopted the separate-department concept. The argument for the separate drafting department is that by having a nucleus of experienced draftsmen, standardization of drafting procedures with its efficiencies can be promoted and such an organization would minimize the amount of time that creative engineering talent would otherwise spend in routine drafting efforts.

From Figure 12.1, it can be seen that the drafting department of Acmen Electronics is organized as a separate organizational unit under the cognizance of the project design engineer. The lines of communication would be routed from the design engineering groups to the chief draftsman via the project design engineer.

The communication between the design engineering groups and the drafting personnel is much less formal than in other parts of the organization because of the unique form of information that is transmitted. The draftsman is in effect translating into formal engineering drawings the ideas and concepts that the design engineer creates. The documentation by the design engineer usually exists in the form of sketches and notes which must be verbally explained to the individual draftsman doing the work.

However, in spite of the nature of the task, formal instructions must be initiated at the start of any drafting requirement for planning and scheduling purposes. Therefore the engineering department will use some form of Drafting Order to be issued to the chief draftsman. The Drafting Order will contain pertinent information that the design engineer must document in order that the drafting department may fully understand and interpret what must be contained in the drawings. The Drafting Order will include the following information: description of task, supplementary sketches and other documents, budget hours to complete, and date drawings are required.

After receiving the Drafting Order, the chief draftsman can schedule the work and provide the required detailed information which would permit the assigned draftsman to accomplish the task.

Any revision or change in the design must be transmitted as an instruction to the chief draftsman and would be contained in a Drafting Order Change document.

14.7 Purchasing Instructions

A requirement for purchasing is originated by the project design group, submitted to the project engineer for approval, and formally issued as a requisition for purchase to be executed by the project purchasing agent.

The factors that must be considered in reaching a decision as to "make or buy" a component or subsystem were discussed in Chapter 12. When the decision to buy is made, the requisition, which contains complete detailed information with specifications, quantities, required date, suggested vendors, etc., is routed to the reliability-control department to verify that the item will meet the reliability standards for its application. The requisition is then routed to the project purchasing agent.

Once the purchase order is placed, copies of the document which give all the pertinent information relating to the purchase are sent to the project engineer and the project design engineer. A check must be made by the design engineering department to verify that the part being ordered does, in fact, meet the specified engineering and reliability requirements. Subsequent to the placing of the purchase order, events may occur relating to changes in delivery, design, etc., which require revisions to the purchase order; and associated with these revisions, changes in the communication channels which would duplicate those described for the initial purchase order are necessary.

14.8 Production Instructions

The monitoring of the production and assembly of a complex piece of equipment involves many disciplines and controls. The project engineer would become involved in production details only when an unusual problem arises which might jeopardize the entire project. Basically, project engineers are only concerned with the production schedule, quality, and cost, and it is these factors which form the areas of reports to them.

The initial instructions issued to the project production coordinator are general in nature and in essence direct that the necessary planning be made in accordance with the analysis and information that was derived and used for the technical proposal that resulted in the contract award. The document used to convey this instruction is sometimes referred to as the Project Production Order (PPO). In the case of the landmass simulator the Production Order would be supplemented by the specifications, copies of the technical proposal, and other related documents to provide all the necessary information.

The project production coordinator would then break down the project into the various tasks, such as sheet metal work, cable assemblies, and machining operations, and lay out an estimated schedule based on the date that inputs from engineering are expected.

When the engineering design is established and the drafting department has completed the engineering and manufacturing drawings of a particular component or assembly, the information is incorporated into a Production Work Order and transmitted to the production coordinator for further transmission to the particular department that will fabricate and assemble the part.

The communication of the Production Work Order would be routed through the project engineer who authorizes the work and verifies that the indicated schedule of work does conform to the master schedule of the project and that tolerances and quality requirements are satisfactory. Periodically, the project production coordinator will transmit a status report relating the information discussed in Section 14.2 to the project engineer.

14.9 Schedule and Cost Coordinator Communications

From the project organization chart shown in Figure 12.1, it is noted that the schedule and cost coordinator does not exist in the direct line of authority but functions in a staff capacity to the project engineer. The primary function of the coodinator is to gather information as to the project cost and schedule status, compare the information with the planned and budgeted data, and submit status reports to the project engineer.

Initially, the project engineer would submit the detailed cost and schedule requirements including the breakdown of the elements of such requirements to the coordinator. The project engineer must break the information down into categories which will be consistent with the information to be received in the future, reflecting the experienced cost and schedule figures. For example, the estimate of costs for the landmass simulator was described in Chapter 6, and the costs were derived by dividing up the project in accordance with the design detail shown in Figure 3.1. Thus, the project engineer has a detailed cost breakdown based on different systems such as the Shadow Computer shown in Table 6.1. It may be that the actual effort may not be accomplished in accordance with the breakdown as evolved during estimating, and therefore the correlation of actual cost elements with original estimated cost elements would not be feasible unless the cost estimates were reoriented. In further explanation of this point, in Table 6.1, assume that the Shadow Start and Shadow Stop portion of the Shadow Computer are subcontracted to an outside company. The project engineer would then have to shift the costs for these two circuits to purchased materials and adjust the estimated Shadow Computer figures accordingly.

The schedule and cost coordinator would receive periodic status reports from the project design engineer, project production manager, project purchasing agent, and project quality coordinator. The information so received would then be presented in a formal narrative report and would be graphically depicted in such displays as the Cost Prediction Report (Figure 11.5), Schedule Prediction Report (Figure 11.6), Manpower Loading Display (Figure 11.4), LOB Report (Figures 11.7 and 11.8), and any other similar documents that are deemed necessary by the project engineer.

In addition to providing the statistical information discussed above, the coordinator must provide information relative to areas where difficulties exist and describe what factors are responsible for the troubles. Also, the coordinator, wherever possible, would provide remedial actions as suggested by the department heads. In this manner, the project engineer would have a complete briefing on the project and be in a position to make a comprehensive, intelligent, and constructive report to higher levels of management.

14.10   Communication with the Customer

The project engineer is also the focal point for communications outside of the organization. In particular, this is the individual to whom the customer directs any inquiries, instructions, or suggestions relating to the project. Very often, a company employs a marketing director whose primary functions are maintaining good relations with customers and functioning as an intelligence source for obtaining various types of procurement information. Although the marketing director may be the official contact point for a customer, practically all questions are passed on to the project engineer, who provides the answers and usually ends up as the contact.

The major forms of communication between the project engineer or the marketing director and the customer are (1) letters, dispatches, and other forms of the formal written word, (2) telephone conversations, (3) personal meetings relating to the monitoring effort, and (4) program review meetings on the project.

With regard to the written communication, it is mandatory that the project engineer see to it that any communication requiring a reply is promptly answered. Prompt replies to letters and other written communications are vital for the following major reasons:

1. A prompt reply minimizes any delays in the program progress by contributing to prompt decision and action.

2. A communication relating to some point, if not answered before a specific period (usually 30 days), is assumed to constitute concurrence with the point in question, and the addressee, by remaining silent, is assumed to have given tacit approval to the customer's point of view. Contractually speaking, then, the rights of the project engineer will be lost on any issue in a letter which was not answered in time.

Another vital point relates to verbal discussion such as telephone conversations and meetings. Written records of the content of all such communications should be kept. In the case of meetings, minutes usually are published and distributed. For other types of communication, a written record of any conversation is mandatory. If the conversation relates to any decisions, action items, or other significant matter, the project engineer should immediately draft a letter to the other party confirming what was discussed and detailing the agreements and action items.

14.11 Decision Making

The fundamental purpose of providing for channels, methods, and tools of communication is to permit responsible management officials to arrive at proper decisions. At this point, the fundamentals involved in arriving at a decision and how the decision-making function is contingent upon communications will be discussed.

1. Identifying the existence of a problem is usually facilitated by the fact that some objective is not being achieved. For instance, the effort or cost to complete the design for the Shadow Computer may be exceeding the budget. Investigation might reveal that the cause of the problem is due to a particular design problem. It is essential that the problem be defined as precisely as possible.

2. The accumulation and evaluation of the facts related to the problem constitute the second phase in deriving a decision. In the case of the design problem cited above, the specification requirements, the conditions, the specific technical requirements, and the difficulties which hinder the achievement of those requirements must be accumulated and evaluated.

3. Consultation with other individual experts in the area in which the difficulties exist is the third phase of the decision-making pattern. The introduction of outsiders to consider the problem brings into the picture new points of view which may result in a solution that was overlooked by the individuals living with the problem. The least that the outsiders may offer is that the existing design approach provides no solution and an alternative approach must be processed.

4. The fourth phase, suggested above, is to seek alternative approaches if the existing approach is judged to offer no solution. If, in fact, the existing path of action is deemed to be unfeasible, the earlier a new approach is adopted, the less the schedule slippage and wasted effort.

5. The fifth phase is to analyze thoroughly the proposed solution to the existing problem or to evaluate in detail any alternative approach to make certain that whatever course of action is taken, the result will be successful.

6. The last phase is the actual decision itself and action to implement the solution to the problem.

Whereas the mechanics of identifying the problem, gathering information, etc., are essential prerequisites to the decision-making respon-

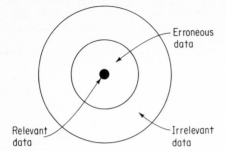

FIG. 14.2   Types of data gathered for consideration of a decision.

sibilities of the project engineer, the actual point of executing the decision requires mental disciplines that should be recognized. A decision must not be influenced by any emotions or personalities. Therefore the data relating to the problem must be screened to establish its relevancy. Figure 14.2 illustrates the types of data that are usually accumulated with regard to a problem and the amount of data that is normally considered relevant. Project engineers must, with complete objectivity, identify and discard the irrelevant and erroneous data so that they can concentrate only on the relevant information as the basis for their decision. It should be noted that if any erroneous or irrelevant information is used as a basis for a decision, the logical conclusion that can be reached is that the decision itself will be in error.

14.12   Summary

The successful monitoring and rendering of decisions on a project is dependent on the effectiveness of the communication system of the program. The project engineer who is the key person in a program is dependent on the prompt receipt of accurate, complete, and precise information.

The two categories of communication are the routine periodic status reports and the special communication initiated by the presence of a problem. In addition to receiving communication from the working levels, the project engineer must transmit information to higher management. The receipt of communication forms the basis for decisions which are translated into instructions on the project.

The management echelons of a company are primarily interested in whether the specification requirements, estimated costs, and contract delivery schedules are being met since the determination of a successful project must meet all three objectives.

The tools of communication vary in detail for different companies but essentially are all designed to convey the general program status, the cost status, and the delivery schedule.

Project engineers, in carrying out their responsibilities, must initiate proper instructions to the various department heads who report to them on the program. The instructions which exist as formal documents must then be broken down into detailed instructions by each of the department heads and transmitted down the line to the individual doing the job.

In addition to the formal documented instruction, there must be verbal communication to convey interpretations and provide feedback as required in order that a project proceed to a successful completion.

Schedule and cost coordinators receive information relating to the project objectives and periodic information reflecting the status of the program at particular points in time. They compare the two categories of information and transmit the results to the project engineer. The information thus conveys the exact status of the program and permits further action as required.

In addition to the internal communication modes, the project engineer must be in touch with the customer regarding aspects of the program. The relation between the project engineer and the customer is by necessity formal and sensitive since whatever may transpire between the two parties can have significant contractual implications.

The communication cycles form the basis for rendering decisions. The fundamental actions involved in arriving at a decision are:

1. Identify the problem.
2. Accumulate and evaluate the facts.
3. Consult with others.
4. Seek alternative approaches as necessary.
5. Analyze all possible solutions.
6. Make a decision and implement it.

PROBLEM

In attempting to meet the very stringent requirements for weight, a structure assembly of complex shape was designed to be fabricated out of magnesium. The machining section of the production department was experiencing unusually high rejection rates on their metal-cutting operation. The basic problem apparently was the tendency of the metal to tear.

The problem was referred to the project engineer for resolution. Analyze and describe the steps to be taken to arrive at a decision for the problem's resolution.

Chapter Fifteen

Engineering the Equipment

15.1 Planning the Design

The receipt of the Task Assignment Order by the project design engineer is the trigger for initiating the official engineering effort on the landmass simulator. The task number will be used throughout the life of the program to identify the project for which work, materials, and parts are expended. The task number is also used for cost control, accounting, and scheduling on the program. The responsibility of the chief project engineer will be to create a design for the simulator based on the approach described in the technical proposal which will (1) meet the specification requirements, (2) embody the least costly approaches, (3) achieve the greatest degree of reliability and standardization, and (4) be completed within the time schedule of the program for the design effort.

The project engineer would consult the project design engineer to establish the master schedule and examine in greater depth the approach for the design effort. Regardless of the project and the technical disciplines involved, the same fundamental planning and scheduling must be thought out for any design engineering effort. For the landmass simulator project, Table 15.1 illustrates the general periods of effort on activities that must be planned. These activities are analogous to those of a PERT program.

**TABLE 15.1   General Schedule of Engineering
Activities for Radar Landmass Simulator**

Period in weeks after contract award	Type of effort
0	Contract award
0–4	Design data received
4–10	System engineering completed
10–24	Design completed
24–26	Components and subsystems established
26–28	Breadboards verified
28–30	Packaging design completed
30–48	Drafting completed
56–65	Integrated test and checkout completed

15.2   Procurement of Data for Design

After receiving the Task Assignment Order, the project design engineer will have the system engineering effort completed and subsequently have each of the engineering group leaders itemize the design approaches for the portion of the RLS for which they have responsibility. The group leaders will also screen and catalog the data which must be transmitted to Acmen Electronics as Government Furnished Property in accordance with the contract terms.

The issue as to what degree of responsibility the government has in assuming the contractual obligation of delivery data as GFP to a contractor is one which is constantly debated. The procurement of data for simulator equipment is particularly critical on projects where the data relates to operational systems which are still under development or which have only recently been perfected. Often the data may not be formally documented and therefore are not available from the usual sources. The design of a piece of equipment such as a simulator would require specific data that must be extracted from a mass of general information pertaining to a larger operational system or complex. The contractor's design engineers assigned to the project would be the most qualified individuals to identify the specific data that are required. Therefore, even though contractors may not be obligated to participate in procuring data, in the interests of expediting the progress of the project, they must identify the areas of data that are required.

If, for any reason, the government failed to provide the data in accordance with the contract schedule and if the contactor could substantiate the out-of-scope effort that was necessary in executing the contract because of delays in the receipt of data, then a justified claim for additional funds and/or an extension of delivery could be made by the contractor. In the case of such a claim, the government would generally require that the contractor cite the specific data that were required and not received on time and document how the delay resulted in the adverse affects on the contract.

The tabulation of the specific data that are required would be accomplished by the individuals assigned to design the various subsystems. The tabulation would be transmitted to the contractor project engineer who in turn would pass the specific data requirements on to the customer's project engineer. If it develops that a particular area of information is not available in its formal state, special arrangements and procedures can be implemented between the project engineers of the government and Acmen to jointly obtain the information and data by discussing the requirements personally with the third party who, for example, might be the design engineers of the company building the AN/APQ-28 Radar System, or they can obtain preliminary information by some other means. The whole point is that the cooperation and best efforts of both the customer and the contractor are essential to overcome problem areas and contribute to the success of the program.

15.3 System Design

The complex equipment of today consists of many subsystems which interact with each other. The product must function within specific tolerance limits and perform to meet established objectives. The interaction among many subsystems requires that the design engineer look at the system as a whole before getting in the detail design of subsystems and their modules. The review of the total system and its analysis has resulted in a category of engineering effort identified as system engineering.

In Chapter 3, the tiers of design detail that were discussed for the RLS represent one of the preliminary steps that is involved in system engineering. In like manner, the procedure described for estimating the costs of the simulator in Chapter 6 utilizes some of the concepts related to systems engineering.

An initial step in implementing the system engineering effort is to review and identify the complete system in terms of the specific parameters. Using the RLS as the case to be demonstrated, the following describes the type of effort for the initial phase:

1. *Reaffirm equipment performance objectives:* The simulator is to be designed to present realistic radar returns of a designed geographic area in real time.

2. *Define limits of equipment performance:* This area would include the

design tolerances of such areas as resolution, etc. Since the upper limit of resolution is specified as 250 feet, it would be of no advantage to spend the effort and money to attain resolution of 150 feet.

3. *Identify constraints on equipment design:* What performance requirements pose a challenge to knowledge and technical resources that are available? For instance, can the small spot size of the FSS be achieved or does the system engineer have to live wih the larger FSS spot size and compensate by using a larger transparency scale or use some sort of optical focus system?

Much of the preliminary work relating to the first steps that the systems engineer would take should be accomplished during the preparation of the proposal as discussed in Chapter 4. A well thought-out technical proposal, updated to reflect any changes that occurred during negotiations, would serve as a valid starting point for the system engineering effort.

The second phase of engineering the system is to analyze and define the subsystems of the equipment. This requires that the subsystem functions be identified before the design is pursued.

After the subsystems of the equipment to be designed have been analyzed and defined, the following includes some of the major points that must be considered and resolved before the system design is completed:

1. Interaction among subsystems.
2. Compatibility among subsystems.
3. Relevancy of the design features that are to be adopted. (In other words, to what degree will the subsystem and total system design meet the equipment performance requirements?)
4. Effectiveness or practicality of the design. This consideration relates to the cost in money and time that the design entails.

Basically, system engineering can be defined as a management engineering effort for designing a total system in a manner which will provide subsystems compatible with each other and provide a design capable of meeting the equipment performance requirement in the most cost effective manner.

15.4 Block Diagrams

Figure 6.2 shows a block diagram of the Shadow Computer. The block diagram is one of the key tools which the system engineer uses as an aid to visualizing the various subsystems of a complex design and to establishing the flow of signals among the subsystems.

The block diagram which is drawn up for the technical proposal would be used for general planning purposes and for preliminary task assignments in the design areas. Once the contract is a reality, the project engineer must have the proposed design approach reviewed in a much more critical manner and have the system design depicted with a greater depth of detail. The project design engineer would be instructed to have the detailed system block diagrams made with the inputs obtained from the various groups who have been involved in the project, or who are expert in the different disciplines involved.

The creation of these system block diagrams will afford the opportunity to clarify the specific approaches. When the system block diagrams are formally approved by the project engineer, the project design engineer will initiate the Engineering Orders, each with its specific details, and establish the responsibilities of each group leader for meeting the assigned technical task, design criteria, schedule requirement, and budget limitation.

The overall system block diagram which is composed of the diagrams of the various subsystems permits the systems engineer to determine the compatibility among subsystems. If, for instance, the Shadow Start gate shown in Figure 6.2 requires a specific type of input signal derived from the differentiator, then the design of the differentiator must be accomplished to provide the necessary compatibility. The necessity for emphasizing attention to compatibility is particularly important when the various subsystems of a piece of equipment are being designed by different engineering groups. In evolving the system block diagram (and all throughout the subsequent design effort), constant reference must be made to the specification to guarantee that the technical design being evolved will result in a product that conforms to the requirements.

It should be understood that the detailed specification requirements establish what the customer desires and whatever margins are deemed necessary to guarantee the accuracies and equipment acceptability that have already been considered and have been reflected in the specification. It can be assumed that the customer does not desire anything over and above the requirements of the specification. Thus, the practical goal of the contractor is to design a product that meets the minimum requirements of the specification. Anything over and above that would be unnecessarily costly to the contractor and, in some cases, might even be unacceptable to the customer. For example, the detection of landmasses with radar systems will under certain conditions experience peculiar phenomena known as scatter effect. The simulation of this effect involves a complex electronic circuit and the storage of special data in the transparency. The absence of any reference to the scatter effect indicates that the customer does not expect to have the scatter effect incorporated in the design and, in all probability, does not desire the capability. The project engineer and the project design engineer must make certain that the design being evolved reflects precisely what is required by the specifications and that some engineer designing a particular element of the simulator is not interpreting the technical requirements to provide more than is necessary. The means by which such control can be implemented is by the review and interpretation of the block diagram prior to issuing the order to start the subsystem designs.

15.5 Design of the Subsystem

Upon receipt of the EO the individual design engineer will pursue the assigned task. The design effort for any two tasks will be different, and the creative processes of different engineers will differ. However, there are certain fundamental procedures that must be established on any project or in any organization if for no other reason but to afford management a tool to measure progress or determine the status of a task.

The procedures that the project design engineer would establish for each area of design would conform to the following outlines:

1. Translation of the block diagram, specifications, data of the radar system, etc., into criteria for the design

2. Establishment of the equations and mathematical models which reflect the parameters and functions indicated by the block diagram and related data

3. Establishment of a flow diagram (electrical systems), a force diagram (mechanical systems), etc.

4. Determination from the flow or force diagram the minimum number of elements such as servos that can do the job

5. Conversion of the flow diagrams that would be applicable to the RLS into schematics

6. Establishment of the values of elements, selection of components, etc., which would constitute the design of the hardware

7. Check of the finished design to verify whether the functional equations, design criteria, and specification requirements will be met by the design

Although project engineers would not normally become involved in the detailed design effort, they do have the responsibility of working out general design procedures with the project design and system engineers and approving such procedures for the program. They therefore must have a clear concept of what is involved in each procedure and what benefits they plan to realize from each step of any procedure that is established. Generally speaking, a positive benefit should be realized from any particular document, report, or procedure that is implemented in a program in order to justify its existence.

15.6 Design Check for Standardization

Practically all organizations which become involved with developmental work have a special group whose main function is to analyze the design of a system and implement standardization wherever possible. The average engineer engaged in the design of a system is concentrating on creating a system that will satisfy the requirements for operation and, if some special or nonstandard element will result in a workable system, will generally adopt the element for the design.

Because of this basic trait of human nature to use any acceptable means to accomplish a task in the easiest and fastest manner, design engineers will resist any suggestion or instruction to revise their design to accommodate a standard part or element. Many organizations have a special group whose function is to objectively analyze the design and determine whether a nonstandard part can be replaced with a standard part without requiring extensive revision of the system parameters.

Some of the more progressive companies who have oriented themselves toward standardization train and instruct their engineers to design their work around existing standard subsystems or modules wherever possible. If the design requires a servo system of a particular response speed, stability, etc., the engineer would select a predesigned servo and adapt its use to the system being designed. However, the design must still be reviewed by some disinterested party to verify whether the maximum effort has been made by the designer to use standard subsystems and components.

Historically, standardization of parts has been implemented in industry as an inherent facet of mass production. Industrial concerns, competing for markets for their products, recognized that standardization of parts enhanced their ability to produce cheaply and permit replacement of worn components by the user with a minimum of effort.

More recently, the United States government, which is the largest purchaser of equipment in the world, found that because of a passive standardization program, it was necessary to support a fantastic number of spare parts of every conceivable type, with each part bearing a different catalog number. An exhaustive study revealed that two companies using an identical part in equipment being furnished to the government assigned different part numbers, therefore requiring that two sets of spares be maintained in a supply depot when one set would have sufficed. Further, companies had a tendency to design special components for use in equipment when a standard unit readily available in the market or in existence in the government supply system would be completely adequate.

In an effort to simplify the support of equipment furnished, the government drafted several different specifications for use in contracts which set forth requirements to effect standardization of equipment components. One commonly used specification sets forth five groups of requirements for standardization. The minimum standardization requires that a contractor select parts which adhere to the company specification or standards. The maximum standardization requires that the company use parts which conform to government-wide standards such as JAN (Joint Army Navy) standards. The government standards specification requires that the contractor select components that conform to the most universal or group I standardization requirement. If a required component complying with group I standard does not exist, then the contractor would have to investigate the possibility of using a group II component, and so on. If a required component is so unique that it cannot meet the requirements of any of the five groups of standards, then special permission must be received from the government to use the special nonstandard component.

Project engineers, who are primarily responsible for the standardization program, would have to coordinate their project with the company standards group or committee. One of the functions of the standards group is to compile a library of standard components, modules, subsystems, and equipment used by the company in its operation. The standards group would be responsible for analyzing the design of any system to determine whether the standards requirement has been met as previously discussed. The project engineer would have the final decision whether to use a standard component or subsystem in a design in the event of any conflict between the standards and design departments.

15.7 Reliability

The reliability of operation of a system can only be defined in general terms. A more detailed discussion of the concepts of reliability will be discussed in a later chapter. There is no standard definition that exists since different systems may demand different degrees of reliability. The reliability requirements for the RLS are expressed in the specification excerpt discussed in Chapter 2.

The most common cause of poor reliability in equipment is the failure of electrical and electronic components. The major cause of failure of such components is heat. Life tests have been conducted on practically every conceivable type of electrical and electronic component, and the results are reflected in curves showing the life expectancy of components as a function of environmental or ambient temperature. In general, it can be shown that the life expectancy of an electronic or electrical unit will decrease drastically as the ambient temperature and/or the operating temperature rises. The design engineer must design equipment so that the ambient temperature is maintained at reasonable values and so that sensitive elements are not overloaded and are provided with adequate heat sinks.

The design engineer charged with the packaging and layout of the simulator must establish the most effective means of carrying off the heat generated by the electrical equipment. This responsibility is generally given to the mechanical engineer. A deficient design will result in hot areas in the equipment in which the ventilating system of the equipment fails to carry off the heat as fast as it is generated. The electrical or electronic components operating in the hot area are the first units to fail.

The load transmitting shafts, gears, bearings, and other mechanical elements must be designed with an adequate margin of safety to minimize their chance of failure. Although most of the failures of rotating or moving parts that might occur are due to misalignment which would be the result of poor workmanship during assembly, the engineer can minimize the possibility of misalignment occurring by creating a design in which the number of moving parts, gears, and other mechanical components is kept at a bare minimum. Also, the layout to facilitate ease of assembly and later maintenance would contribute to minimizing misalignment problems.

The engineer assigned to review the design of reliability would make an analysis, taking into consideration the various factors noted above. In essence, the key to achieving a reliable design is to keep the design conservative.

15.8   Breadboards

A breadboard is a temporary assembly of electronic and electro-mechanical components which conforms to the engineer's design of a particular system and serves as a means of verifying the design in question. The design engineer involved supervises the assembly of the breadboard and conducts the tests that are necessary.

In order to realize the maximum advantage from the breadboarding program, the components used should be the same as those to be used in the ultimate equipment. The inputs to the breadboard will generally be synthetic but will duplicate to the greatest degree possible the inputs to be experienced by the final equipment. If the performance of the breadboard design does not meet the design criteria, modification such as substituting components with different characteristics is necessary until the satisfactory breadboard performance is achieved. The use of breadboard techniques permits the engineer to make revisions at an early stage in the design effort at a minimum cost and expenditure of time and thereby to eliminate the necessity of making costly design revisions after the equipment is fabricated. For highly complex systems in which there would be so many parameters and variables, both known and unknown, that it would be extremely difficult to consider all such variables, a complete system would be breadboarded and verified before the design is adopted for production.

In the case of the RLS, the project design engineer would determine which subsystems are unique to the company and therefore should be verified by breadboarding prior to initiating the drafting and fabrication efforts. Since Acmen Electronics has a minimum of experience in several areas of the simulator design such as the Shadow Computer, directivity effects generation, and beam control system for the FSS, it would be logical to include these designs in the breadboard phase for verification.

15.9 Packaging Design

For electronic equipment, the mechanical engineer is primarily responsible for meeting the specification requirements for packaging which includes the shape, layout, size, and general configuration of the equipment. The mechanical engineer must take into consideration such factors as ventilation of the equipment, accessibility of modules for maintenance, and total weight.

The complexity of modern equipment (especially military equipment) would result in hardware prohibitive in size and weight if design techniques and components as recent as 15 years ago were to be used. Because military equipment must be extremely compact in size and light in weight, so-called subminiaturization programs have been undertaken, and techniques using printed circuits, solid state modules, etc., have resulted in highly dense designs and compact units. The limitations of size and weight on the RLS are moderate so that extreme measures for dense packaging are unnecessary. However, the mechanical engineer must complete a design which will be compatible with the specification requirements. If, for instance, the design engineer had specified a series of 3-inch potentiometers which could not be incorporated physically into the system and still meet the specification requirements for size and/or weight, the mechanical engineer would indicate the difficulties and the design engineer would have to design the system using a smaller series of potentiometers.

In addition to the responsibilities for providing equipment that meets the physical requirements for size, shape, and weight, the mechanical engineer is responsible for providing equipment that embodies adequate ventilating features to prevent an excessive accumulation of heat that is generated by the electrical components. The life of components that are subjected to the high ambient temperatures will be significantly shortened. The ability of the equipment to maintain low temperatures is one of the main factors that enhances reliability.

Another responsibility of the mechanical engineer is to design the mechanical structure and enclosure so that the specified environmental tests for shock, vibration, moisture, etc., will be met. The mechanical engineer must not only establish a structural design that will be adequate but must check the major components of the other design areas such as the electromechanical elements to verify the ability of the components to meet the environmental tests.

15.10   Drafting

The drafting effort is initiated when the design of a particular subsystem is established and constitutes what is essentially a formalization of the creation of the design engineer. The drawings which are produced by the drafting department should describe all the detail so that another person qualified to do the work could convert the drawings into operable systems. Because of the amount of detail that must be incorporated in the drawings, a close personal liaison must be established and maintained between the design engineer and those doing the drafting. It would be next to impossible to convey any but the broad description of what is to be incorporated in the drawings by the Drafting Work Order.

There are two basic types of drawings that are generally required. They are so-called engineering-type drawings and manufacturing drawings.

The engineering drawings depict the design details of electrical circuits and similar details that are necessary for the maintenance and modification of the equipment after its delivery. These drawings are generally used in conjunction with the maintenance and operation manuals for the support of the equipment in the field.

The manufacturing drawings show all the construction details of the equipment as well as the design details contained in the engineering drawings. These drawings are provided to the manufacturing department for fabrication and assembly of the equipment. Quite often, the customer will use the manufacturing drawings to procure additional identical units of the equipment if the requirement should develop. In such an eventuality, the drawings are made available to all qualified offerors, and procurement on the basis of a firm fixed-price contract is solicited.

Theoretically, the manufacturing drawings should enable a qualified company to build the equipment with little or no engineering effort. For complex equipment, the idealistic objective is rarely realized since, when one considers that perhaps a thousand drawings may be involved, it would be almost statistically impossible to have the drawings free from any errors or omissions. Procurements based on such premises invariably result in contractual controversies in which the contractor might submit claims against the buyer because of difficulties and delays encountered due to errors in drawings furnished by the procuring agency. Recognizing this potential source of difficulty, several procurement agencies have protected themselves by incorporating language in the contract which requires all offerors to inspect and analyze the drawings prior to submitting their bid and to provide in the bid price enough of a contingency factor for any effort necessary to correct drawing errors and omissions. The contractor thus must provide equipment which will operate as specified in spite of possible drawing deficiencies. This practice has benefited both the buyer and seller. The buyer has eliminated the cloud of potential claims on the procurement. In a similar manner, any marginally qualified company would hesitate to accept the risk imposed due to the responsibility it must assume regarding the drawings.

Because drafting is a time-consuming, meticulous effort, close planning and scheduling must be made to minimize the possibility of exceeding the allotted time in the overall project schedule. Therefore, as soon as the general design and the scope of effort are established, the project design engineer will issue to the chief draftsman the Drafting Order which will be used as a basis for planning and scheduling. The Drafting Order will include a description of the type of equipment to be designed, the number of subsystems comprising the equipment, the time schedule showing when each of the subsystem designs will be completed, the estimated number of drawings involved, the list of detail specifications on format, and other pertinent information to which the drawing must conform.

When the engineering design releases are received, the draftsman will initiate the work and will maintain the close liaison with the design engineer. The type of information that will be forwarded to the draftsman will be sketches, handwritten notes, and other informal documents which reflect the work and creation of the designer. These documents must be elaborated on and explained to the draftsman so that they can be translated into the formal engineering and manufacturing drawings as required.

In a development program, there are bound to be revisions in the design which must be reflected in the drawings. Any revision that is significant or that may affect the drafting completion date for the particular draftsman must be documented in a Drafting Order revision. The documentation is necessary because of the informal personal liaison that exists between the individual design engineer and the draftsman and that may result in a situation where significant design changes are directed by the design engineer which may not be approved by the project design engineer and which may jeopardize the design of the project system. In other words, where circumstances dictate that informal communication in a particular area would benefit the program, the entire organization of the program can break down if the informal communication in the area in question is permitted to get out of hand.

Drafting is one type of work which has not lent itself to any significant degree of automation and still constitutes a time-consuming, tedious hand operation. There have been some attempts to eliminate the necessity of redrawing circuits, structures, or modules which are standard to a company and which are used repeatedly on different types of equipment. An example of such a module would be an amplifier. In such cases, means and techniques have been developed where the module circuit drawing is available as a transcription which is more or less affixed to the work being processed by the draftsman, and its use eliminates the necessity of drafting a complex schematic or structure. These techniques, many of which are patented, are available and their use can result in considerable savings in time and cost.

15.11 Testing

Various types of verification tests are performed as deemed necessary and where possible. The breadboard program previously mentioned is one phase of testing. There are many other phases, practically all of which are under the cognizance of the engineering department.

The test program on a development prototype piece of equipment such as the RLS is one of the engineering functions and must be carefully planned and executed. The details of the test and checkout phase of a project will be covered in a later chapter.

15.12 Value Analysis

A relatively new facet of procurement has been derived which is called value analysis or, in many instances, value engineering. The objective of value analysis is to provide an incentive to contractors to analyze the product for which they have a contract to determine whether a change in design, material, process, or other parameter of the equipment would result in a significant saving in cost but not compromise the utilization capability of the product. The government and other large organizations who engage in procurement to any degree recognize the human tendency to specify and purchase a particular item over and over again without giving any thought to the possibility that significant savings could be realized by using a substitute. A classic example of value analysis in practice occurred in connection with the procurement by the United States Navy of a flexible rope which is snapped between two stanchions at the top of ladders and passageways. The material traditionally used was a special manila hemp with hand-stitched covering of different fabrics. As a result of a value analysis, the substitution of a chain was adopted at a fraction of the cost of the fabric rope.

Although the above example may constitute an overly obvious example of savings accomplished by value analysis, it does bring out the point that an open and inquisitive approach to a situation can result in seemingly obvious alterations to existing approaches.

The achievement of savings by value analysis of complex equipment is much more subtle and requires a much greater exercise of creative thinking.

Contracts which incorporate a value analysis provision embody a sharing arrangement for any cost savings that may be derived from a successful study in this area. Value analysis is generally used in conjunction with a multitude of production items and has very little application in prototype development units. However, project engineers should be cognizant of the potential for profit to be derived from a value analysis program so that they can implement the effort where applicable.

15.13   Summary

The success of a project which requires any degree of research is dependent primarily on the effectiveness of the engineering effort. Therefore, engineering must be planned and controlled to provide a design that meets the specification, schedule, and budget requirements of the contract.

The individual charged with the responsibility for engineering is the project design engineer, who is directly responsible to the project engineer in the illustrative case of the landmass simulator.

The procurement and gathering of data is the first function that must be performed as a prerequisite to the equipment design. Part of the data-gathering chore is to screen the information to derive what is pertinent and to identify and procure what data are required which may not be formally documented.

A complex system comprising several subsystems requires that a system engineering effort be applied in order to achieve compatibility among subsystems and assure that the overall performance and cost objectives of the equipment will be achieved.

The block diagram is the fundamental engineering drawing which indicates the general design approach to be taken. The diagram also indicates the logical breakdown of the various engineering disciplines that would be required and is referred to in assigning various design tasks.

Standardization of subsystem and designing for reliability are two major requirements for modern equipment and are generally specified for military equipment. The responsibility for verifying whether these requirements are met in the design is usually vested in a separate reliability and standardization engineering department.

Breadboards are extensively used to verify a subsystem design and to permit design revisions that might be necessary to guarantee that the subsystem will ultimately perform as required.

Packaging is a function of the mechanical engineer and involves efficient arranging of components, providing adequate ventilation, and designing a structure and housing that will not only meet the size and weight requirements of the specification but will also permit the passing of the environmental tests.

The chief draftsman is responsible for producing the drawings for the equipment being designed and built. There are generally two types of drawings: engineering and manufacturing. The drawings must generally be created from rough sketches and other informal documents which the design engineer produces. Therefore, a close liaison must be established between the design engineer and the draftsman doing the work in order that the finished drawing accurately reflects the design and construction of the equipment.

PROBLEM

The ABC Corporation has a contract for the design and manufacture of an automatic motor-driven pumping unit for use in a sewage treatment plant. The system is comprised of the following major subsystems: (1) water-level sensor, (2) motor speed regulator, (3) power-switching equipment, (4) motor, and (5) pump.

The incoming sewage is fed into a "wetwell" which is a large, deep concrete-lined well used to store the fluid. The speed of the motor-driven pump must be automatically controlled to keep the water level in the wet well at a constant depth. In the event of a failure of the control system, an automatic alarm sounds when the water level exceeds a certain height and provision must be made for the operation of the motor manually. The general block diagram of the system is shown below:

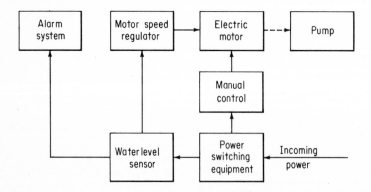

The system must be delivered in 1 year from the date of contract. The following general schedule is applicable:

Engineering design and drafting .... 7 months
Manufacturing .... 4 months
Test and checkout .... 1 month

Lay out the plan and schedule that the project design engineer and the project engineer would evolve for the program.

Chapter Sixteen

Digital Computer Applications

16.1 Computer Introduction

The technology of digital computers has, in the last 15 years, been characterized by its phenomenal advances and improvements. As one generation of computer hits the market, new generations are being developed and tested for introduction to industry. Examples of some of the different types of technological advances that have characterized different operations of computers in the past include the transition from integrated circuits to chips as well as new types of memory module. These types of innovations have resulted in computers that are more compact, faster, more flexible, and less expensive than preceding generations. Because of the dynamic nature of the computer technology, applications that were judged not feasible in the past suddenly can be handled more effectively and cheaper by a general purpose computer of the latest generation.

Individuals following a professional career must constantly update their knowledge to keep pace with the most recent advances—whether it be in the medical, legal, financial, or any other field. Keeping knowledgeable of the advances in the technical field is particularly important for project engineers and their organization. Any company attempting to design and build an electronic system using the technology of 5 years ago would fall far short of achieving the most effective design that could otherwise be realized.

The first generation of digital computers that were available were relatively slow, had limited memory storage capacity, and were generally suited only for applications where total time for completing a computation was not restricted. For instance, the processing of inventory records was not bound to any time constraint so that performing such a function was superior to previous methods for that type of function. However, for applications which required that a calculation be performed in real time—real time meaning that the calculation results be provided almost instantaneously—the general purpose digital computers were generally not fast enough. For instance, in simulating the radar returns that are displayed to a trainee as the radar beam sweeps across a geographical area, the terrain characteristics have to be displayed practically instantaneously—as they occur in operational situation. Therefore, the calculations have to be accomplished in such a short time so that any time lag between the action of the trainee (control manipulation, etc.) and the results of such actions (display presentations, etc.) would not be discernible.

The analog computers, traditionally used for real-time calculations, performed its role well in simulators and similar applications. The inherent design of the analog computer incorporated elements that were precalculated, e.g., especially designed potentiometers, which generated continuous solutions to a particular input and therefore were natural real-time computers. One major shortcoming of the analog computer was its inflexibility for changes. A new characteristic to be simulated required the designing and installing of new elements in the computer. The industry, involved with solving problems in real time, had to live with inherent disadvantages of the analog computer but waited with keen anticipation for the day when a new generation of digital computers would be feasible for real-time applications. For many real-time applications, the latest digital computer design is fast and flexible enough to perform acceptably.

The digital computer is basically a machine consisting of elements or switches that are either on or off. Each element or combination of elements can be switched with lightning speed —at a rate of 1 million times a second or better. However, the calculations must be performed sequentially and in some applications the total time to perform the sequential operations may not be acceptable.

The fundamental parameters that must be considered by the project engineer prior to deciding whether the use of a digital computer is feasible for a particular application are memory capacity, word length, and speed. It is not necessary nor expected that project engineers be expert in computer programming, mathematical modeling, or other specialized digital computer disciplines; but in order to make managerial decisions regarding the use of a digital computer on their project, they must be competent to translate their computer requirements into those three aforementioned parameters of memory, word length, and speed.

16.2 Computer Characteristics

The digital computer can be considered analogous to a worker on an assembly line performing the tasks of processing and assembling a number of metal components. A summary of the tasks and the time periods for performing each operation or function is provided in Table 16.1.

In the above example, the worker can be considered the computer and the operation functions constitute the computer program. The time required to perform each of the functions is analogous to the computer speed. The speed of the assembly line which feeds in a new set of components is the time constraint on the computer. The memory bank would be the assemblies and nuts and bolts that are fed down the conveyor belt. The word size can be considered the jig that governs the tolerances of the assembled article.

In the example discussed the time constraint is 1 minute, which is established by the speed of the continuous belt feeding new components to be assembled at the rate of one set every 60 seconds. As noted, all the assembly-line functions can be completed in 57 seconds.

Since the computer speed is the major parameter that must be considered in computer applications, the timing issue will be the major consideration in this example. Assume that the worker functions were revised and required the drilling of four holes in the assembly. The worker would then be reprogrammed as shown in Table 16.2.

TABLE 16.1 Functions of Assembly Line Worker

Operation number	Function	Operations per cycle (60 seconds)	Operation time, seconds	Total operation time, seconds
1	Fit pieces	1	7	7
2	Drill two holes	2	10	20
3	Fit two bolts	2	7	14
4	Fit two nuts	2	8	16
5	Spare			3
	Total seconds/cycle			60

TABLE 16.2   Revised Function of Assembly Line Worker

Operation number	Function	Operations per cycle (60 seconds)	Operation time, seconds	Total operation time, seconds
1	Fit pieces	1	7	7
2	Drill four holes	4	10	40
3	Fit four bolts	4	7	28
4	Fit four nuts	4	8	32
	Total seconds/cycle			110

Since the maximum time that the worker has available for the performance of the functions is 60 seconds, the 110 seconds required for the reprogrammed tasks mean that the functions cannot be completed within the allotted cycle time. What would happen, of course, is that the components coming down the assembly line at the rate of one set each 60 seconds would pile up. In like manner, a digital computer which cannot perform all its functions within the allotted time cycle would be faced with the same dilemma as the worker, and as a result the computer would halt or stop functioning.

To pursue the analogy of the assembly-line worker and the digital computer, the timing problem that is described can be relieved in a number of ways. Some of the more obvious are as follows:

1. Replace the worker. If a faster worker can perform each of the functions listed in Table 16.2 in one-half the noted time, the total task can be performed in less than the required 60 seconds. In the analogous computer, a faster computer could solve the problem.

2. Improve the program. It might be feasible to reprogram the worker to simultaneously drill the four holes and fit and secure the bolts so that the total time is reduced below the 60-second limit. Relating back to the computer analogy, a more efficient program could improve the timing problem to the point where the previously unacceptable computer can be made to perform satisfactorily.

3. Extend the cycle time. The entire operation can be reviewed to determine if the total cycle time can be extended from 60 seconds to 120 seconds. In the case of the assembly-line worker, the cycle time would be achieved by slowing down the assembly belt so that the components are fed to the worker at a rate of one set every 120 seconds. However, the option of extending the cycle time of a computer, even though possible, would not normally be done because of the degradation of the computed results.

4. Replace the existing worker who is skilled in the multidisciplines

of drilling and utilize two workers with specialized skills, one of whom is skilled to drill and the other to fit the pieces. This option would be comparable to replacing a large computer with two minicomputers for processing specific information.

The ability of a computer to perform its functions within the specified time limit is a major consideration of computer application and selection. The size of the computer data storage and memory is another consideration. Again, the improved technology resulting in high-density memory storage devices, the development of high-speed addressing techniques, and other similar innovations have all but eliminated size of memory as a major problem for computer applications.

Another parameter relating to computer application is word length. The computer word length is expressed as the number of characters (bits) which occupy one storage location and is treated as a unit. The more commonly used general purpose computers are 16-, 24-, or 32-bit machines. A 16-bit computer can count up to over 64,000 and therefore establishes a basic accuracy of 1 in 64,000. A 24-bit computer can provide a basic accuracy 1 in 16,000,000. By adopting special number systems, making use of multiple precision techniques, or adopting other special procedures, the accuracy capability of a computer can be improved. However, the project engineer must consider that in utilizing special techniques in order to achieve special performance capabilities, such as greater accuracies, price must be paid because of greater demands on computer time and/or memory.

In an actual computer application, all bits of a word may not be available for calculation accuracies. For instance, 2 or 3 bits might be required for addressing (memory location, etc.), 1 bit for parity check, 6 bits for designation plus or minus, and so forth. Generally, most of the housekeeping functions noted above can be formatted so as not to compromise the basic accuracy capabilities of the computer except for the plus or minus sign bit which is almost always required.

Another factor that must be considered is that in most applications, the theoretical accuracy relating to the number of bits available for calculations may not be achieved because of the necessity to round off numbers. For example, assume that 8 bits of a computer are scaled to provide an accuracy to the least significant number of 1 and that the problem requires the following four numbers to be added and then multiplied by 6.7:

Actual operation	*Computer operation*
7.8	8.0
9.2	9.0
8.6	9.0
3.5	4.0
29.1	30.0

The computer addition is immediately faced with an error of 0.90. Performing the computer multiplication would provide the following:

Actual number	*Computer number*
29.1	30.0
× 6.7	× 7.0
194.97	210.9

What is demonstrated is that the computer number word must be properly scaled and adequate word length must be considered for any application.

Therefore, in considering the application of a digital computer, the project engineer must have a thorough analysis made and must appreciate the major factors which will affect the apparent capabilities of the digital computer.

16.3   Analysis of Computer Functions

To return to the case procurement of this text, the first parameter to be considered for the radar landmass application is the amount of memory that the digital computer must possess. The total land area to be simulated (elevation, reflectivity, and other characteristic features) is identified in the trainer specification as 1,500 by 800 nautical miles. The specification also establishes that the display must be able to resolve or distinguish objects as small as 250 feet on a side. The total area can thus be divided into grids of 250 by 250 feet which total 6.9×10^8 grid areas. Based on the above, it might be deducted that the computer must have a memory capable of storing at least 6.9×10^8 words. This would be correct if no two grid areas were alike and therefore required a unique word to describe the terrain and other characteristics of each individual area. Since large numbers of grid areas have identical characteristics, the format of a single word could, for instance, describe a multitude of areas. In the case of a desert terrain millions of areas would be identical and therefore only one computer memory location would be required to describe such a large number of areas. At any rate, the project engineer's analysis would establish that the memory capacity of the computer would be significantly less than 6.9×10^8 words and well within the computer technology.

The second parameter that must be considered is the word length that the computer must possess. The more common general purpose computers have word lengths of 16, 24, or 32 bits. One approach that the project engineer might take for establishing the computer word length is to analyze the word format to establish the most demanding word requirement.

The project engineer's analysis of the problem might indicate that the greatest accuracy demand on the computer is that the simulated aircraft be pinpointed to within 100 yards or 300 feet at any instant. The 1,500 by 800 mile total area can be represented as a grid of lines spaced at 100-yard intervals. The grid would consist of 30,000 lines and 16,000 lines in the X and Y coordinates. Therefore the computer must be able to identify the more demanding parameter comprising each of the 30,000 lines in the X coordinate. A 15-bit word can count up to 32,767. Allowing 1 bit for sign check, a 16-bit computer would be adequate for the radar landmass application. Figure 16.1 provides some of the word formats that the project engineer would sketch when considering the computer word size that might be required.

The project engineer would be required to make a similar analysis for other calculations to make certain that there are no other required calculations that may be beyond the accuracy capability of the computers under consideration. Some of the types of computations for the RLS would include radar sweep, calculation for terrain contours, generation of shadow effects, etc. For purposes of this discussion, it can be established that the calculation to establish the X and Y axis aircraft position constitutes the most demanding requirement for computer computation.

The conclusion that would be reached by the project engineer is that a computer having a word length less than 16 bits would prove to be unacceptable and a larger word length computer such as 24 bits would

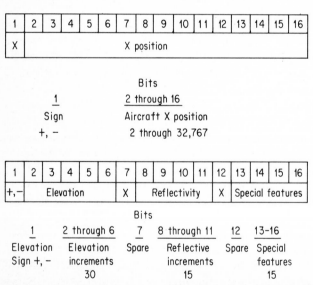

FIG. 16.1 Examples of computer word format for Radar Landmass Simulation, x-axis aircraft position.

TABLE 16.3   Summary of Timing Functions for Radar Landmass Simulation

Function	Memory instruc- tions	Instruction for 4 second cycle				Total in- structions per cycle
		4/cycle	3/cycle	2/cycle	1/cycle	
Area characteristics	1,600,000				1,600,000	1,600,000
Aircraft flight	15,000	15,000				60,000
Radar sweep	5,000		2,000		3,000	9,000
Terrain contours	500,000				500,000	500,000
Shadows	500,000				500,000	70,000
Other	100,000	25,000	25,000	25,000	25,000	250,000
						2,919,000

$$2.0 \times 10^{-6} \text{ sec/instr} \times 2.919 \times 10^6 \text{ instr/cycle} = 5.838 \text{ sec/cycle}$$

be oversized and not cost effective. A 16-bit machine would be the best selection for the radar landmass application.

The third and, in most cases, the most critical parameter that must be considered for a computer application is timing. If the computer cannot perform its assigned functions within the required time frame it generally would experience a halt, or cease to continue its functions. Therefore, for applications which demand that data processing (e.g., a computation) be accomplished within a specific time frame, such as in the case of the landmass simulator, a timing analysis of the computer's ability to perform its functions is required.

A review of the AN/APQ-28 radar characteristics reveals that the fastest antenna rotation rate is one 360-degree revolution every 4 seconds. Also, its effective range at aircraft cruising altitude is 30 miles. Therefore at any particular 4-second period the area that must be displayed is that which is contained in a circular area having a radius of 30 miles. The point of this analysis is that the computer, in the 4-second time cycle, need only process data related to the 30-mile circle or approximately 1.6 million 250-foot square areas instead of processing the 691 million areas that would be necessary of the 1,500 by 800-mile total area. A digital computer with a 2-microsecond computational speed would require 3.2 seconds to individually gain access to or retrieve each of the 1.6 million areas under consideration. In addition, the computer must perform other calculations, such as establishing the area being scanned by the rotating antenna, simulating the flight path of the aircraft, and simulating the sweep time. Table 16.3 illustrates a summary of the computer requirements and the conclusion that would be drawn by the project engineer regarding the feasibility of using a digital computer for the radar landmass application. The summary is based on a 2-microsecond machine which the project engineer concluded was the fastest 16-bit computer available for the application.

As shown in the summary provided in Table 16.3, the total number of instructions that must be executed for each 4-second cycle is 2,919,000. Based on a computer capable of executing an instruction in 2 microseconds (often referred to as computer iteration rate), the total time required to process all the instruction required by the RLS is 5,838 seconds. Since a fundamental requirement of the trainer specification is to update all radar information every 4 seconds (based on maximum antenna speed, area to be displayed, aircraft speed, etc.), the use of the 2-microsecond computer is not fast enough (5.838 seconds required versus 4.000 seconds available). The project engineer would thus have to eliminate the technical approach using a digital computer for the radar landmass application.

The above demonstrates the type of analysis that the project engineer would make before deciding whether a general purpose digital computer is feasible for a particular application. If the program for the RLS was not in a competitive procurement situation or if other factors forced further consideration for applying a digital computer to perform the simulator functions, the project engineer could investigate options such as the following which might prove feasible:

1. Use of special programming techniques
2. Design and development of special data storage and retrieval modules
3. Design and development of a special purpose computer specifically for the RLS
4. Utilization of a hybrid system (digital/analog computer system)
5. Other options

It should be noted that a possible digital approach that might prove feasible in spite of the fact that the straightforward use of a general purpose computer encountered an unacceptable timing problem would involve development, extensive design, and possible research—all of which would require expenditures of time and money. In a competitive procurement situation such as encountered with the RLS, a company cannot risk proposing technical approaches on a fixed-price contract that may not be feasible.

16.4 Summary

The three basic parameters that dictate the application of a general purpose digital computer for performing a task are size of memory, word length, and speed.

High-density memory units and innovative techniques for storing and utilizing data are available so that with current technology, memory very seldom poses a problem for applying digital computers.

The number of variables to be handled and the accuracy of computation that is demanded will dictate the word length or number of bits of a digital computer. Computational techniques such as double precision can be used to permit the use of the smaller word length computer. Consideration must be given to housekeeping chores such as parity check, sign, and addressing which require part of the word that might otherwise be thought to be available for the computational functions.

The ability of the computer to perform its computations within an established time cycle is generally the most severe limiting parameter for a digital computer application. An analysis of the timing functions that are required must be performed in order to determine whether a particular computer is fast enough.

PROBLEM

A project engineer is considering the use of a 5-microsecond digital computer for an automobile simulator to train student drivers. Computations must be performed in a ½-second cycle. The number of system functions that must be performed are as follows:

Function	Instructions	Instruction per cycle
Maneuvering	5,000	10/sec
Engine	2,500	2/sec
Visual	4,000	6/sec
Miscellaneous	2,000	1/sec

Performing a timing analysis of the computer application, determine the total time required per cycle and indicate if the 5-microsecond computer is acceptable.

Chapter Seventeen

Reliability and Maintainability

17.1 Background

The technological advances incorporated in the equipment used for military, industrial, and consumer applications have created serious problems of maintaining reliable and stable operation over even nominal periods of time. Because of the demands by different users for highly automated equipment, designers and manufacturers are forced to emphasize reliability for the equipment that they produce.

There are a multitude of factors that enter into the area of reliability. The determination of reliability is in the final analysis dependent on whether a particular component, element, or module is capable of performing its specific function without degradation or failure for a specific period of time under the condition for which it was intended.

Closely akin to the reliability parameter are the factors of maintainability, availability, and effectiveness. Whereas reliability is the prime concern of the user of a piece of equipment, maintainability or the ability to expeditiously make necessary repairs when some failure occurs is of vital concern in many cases. In the same family of parameters there is also the factor of availability, which is a function of reliability and maintainability. In addition, this chapter will address effectivity which is a measure of the combined factors of reliability and availability of a piece of equipment.

17.2   Reliability Concepts

Reliability is a statistical term generally defined as the probability that for a specific period of time and under specified conditions a piece of equipment or a system will perform in a manner that is acceptable. From the above, it should be noted that when reliability is incorporated as a contractual requirement, the terms of time, conditions, and acceptable performance must be expressed quantitatively. A detailed study of reliability involves statistics and mathematical derivations which go beyond what is required by the project engineer. The primary notions of reliability such as Mean Time Between Failures (MTBF) and availability will be treated in this chapter.

The ultimate in reliability is the ability to have the equipment operate as specified without failures over a specified operating period during the life of the equipment. The modern radios, television sets, and steam irons are examples of equipment wherein an unusually high degree of reliability has been realized. The average television set can be turned on at any time of the day and consistently satisfactory performance can be obtained for years without any breakdown. The degree of reliability achieved with consumer products is the result of exhaustive tests and redesign before releasing the product for manufacture. It is interesting to note that because of escalating costs for repair services, many companies are designing their products to achieve a higher degree of maintainability. Examples are the use of throw-away circuit cards for television sets and similar features for other electronic and mechanical devices which are being promoted by various manufacturers.

Time and money preclude the exhaustive testing and redesign for military systems that are possible for a consumer product. If large numbers of a particular system are built, such as a radar set, design and component improvements resulting from field utilization experiences can and often do result in a very reliable system being realized. However, for a prototype run on a small quantity of a particular complex system, practicality dictates that a compromise be made in the specification as regards reliability.

The RLS specification, for example, expresses the reliability and maintainability requirements for the trainer to be capable of meeting an MTBF reliability of 100 hours and an MTR (Mean Time to Repair) maintainability of 1 hour.

Since the reliability parameter is of major consideration and is intimately associated with the design of the equipment, the discussion will deal primarily with the reliability requirement. Once the reliability requirement has been established, the design engineer must convert this requirement into subsystem design criteria and component selection. For example, the question might be raised as to what would be the difference in design approach if the requirement was for a forty-eight-hour continuous operation instead of a twenty-four-hour period.

The answer to such a question would be apparent after consideration is given to the question of reliability. The degree of reliability is ultimately based on the following two major requirements: (1) the ability of the individual components and elements to perform within their individual specified tolerances and conditions for their rated life and (2) the ability of the design of the different subsystems to function in their intended manner under different specified conditions.

In the case of the individual components, their selection is based on the MTBF values. What constitutes an acceptable mean time between failures is dictated by the specified operation time (e.g., twenty-four or forty-eight hours) and a distribution of the total of all the mean time between failures of components in a system or array of equipment. The statistical distribution of the data will result in a probability curve from which a prediction of reliability can be derived.

The other requirement of reliability is how well the design of the subsystem can maintain the specification requirements for performance under the range of conditions to which the equipment would be subjected. This requirement is achieved by networks, feedback circuits, and various other design features that will automatically compensate for adverse effects due to changes in environmental conditions.

The degree of reliability of performance of any newly developed piece of equipment will depend on the period in the life of the equipment when the test for reliability is being made. Figure 17.1 shows a typical reliability curve that one might expect during the three time periods of the life of a system.

The degree of reliability that would be realized during the equipment infancy will be poor due to the fact that the so-called shakedown period exposes major and minor deficiencies that must be corrected. A large percentage of these deficiencies will cause a breakdown or degradation in the operation of the equipment.

After shakedown, the normal degree of reliability of the equipment would be experienced during the productive period which comprises the major portion of the life of the equipment. The reliability requirements and design objectives would be expected to be realized during this period.

At the end of the life of the equipment, obsolescence begins to affect the reliability of performance at an increasing rate. During this period, failures occur due to the literal wearing out of parts and components. In Figure 17.1, the reliability curve for the equipment shows a decreasing period of time between breakdowns of the equipment until a breakdown

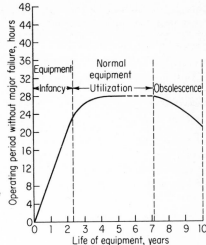

FIG. 17.1 Equipment reliability curve during utilization lifetime.

occurs every 4 hours, which renders the utilization capability of the equipment unacceptable.

At this point, the equipment must be either completely overhauled and refurbished or scrapped. The situation is identical to that which confronts the automobile owner when, after the vehicle has been driven a certain number of miles, breakdowns and subsequent repairs are necessary, and ultimately the frequency of breakdowns renders the automobile useless for the owner. When reliability is discussed, it is to be assumed that the reliability results are derived from operation of the equipment during its normal equipment utilization period or during the center portion of the curve illustrated in Figure 17.1.

17.3 Implementation of Reliability in Design

Prior to implementing the reliability aspects of the design of any system or subsystem, the design and the reliability engineers will jointly make a reliability analysis which will crystallize the environmental and function requirements of the system for which the design engineer has responsibility. Table 17.1 illustrates the reliability analysis for the subtractor circuit of the Shadow Computer of the RLS which has been referred to in previous discussions.

The functional requirements for reliability impose the severest problem for the subsystem design. The method by which an electronic design such as a Shadow Computer circuit is analyzed for reliability will be demonstrated. The principles and logic regarding the approach would be applicable to any type of reliability design, be it mechanical, structural, or electronic.

TABLE 17.1   Reliability Analysis of Subtractor Circuit of Shadow Computer: Radar Landmass Simulator Project #1001

ENVIRONMENTAL REQUIREMENTS:

TEMPERATURE: Normally operated in an ambient environmental temperature of 20°C to 30°C. However, equipment must be designed for operation in temperatures of 10°C to 40°C.

VIBRATION: Equipment must be able to withstand vibration tests specified in attached. Specification MIL-T-17113.

SHOCK: Equipment must be able to withstand shock tests specified in MIL-T-17113.

HUMIDITY: Per MIL-T-17113.

DUTY: Normally, a maximum of 10 hours on, and 14 hours off. Equipment must be able to operate continuously without a major breakdown for 24 hours. Equipment to be capable of utilization over a period of 10 years.

FUNCTIONAL REQUIREMENTS:

INPUT SIGNALS: Input signals to be between 0 and 10 volts D.C. System shall be able to sense difference in voltage inputs of 0.2 volts.

OUTPUT SIGNALS: Output signals to be between 0 and 10 volts D.C. System shall be able to maintain an accuracy of plus minus 3% of calculated value.

STABILITY: Output signals shall be maintained within plus or minus 2% of a particular setting over the entire range of temperature and for a period of 24 hours.

In drawing up a circuit design for reliability, the main problem is to meet the requirements for accuracy and stability as the equipment performs over a period of time. The project engineer is particularly interested in the reliability of the equipment as the temperature is varied from one extreme to the other. If the project involved a structure such as a bridge or a mechanical design such as a machine, reliability or operation over a range of temperatures would involve selecting the proper metals for the members or parts and providing for adequate expansion and load-carrying capacity over the extremes of temperature.

For an electronic circuit design, temperature variation would be handled by properly selected components and designing self-compensation circuits usually taking the form of feedback networks. Most electrical or electronic components change values as their temperature changes. A resistor, when initially energized, will experience a certain value, but over a period of time during operation of the equipment, its temperature will change as well as its value. If the function of the resistor in the circuit is critical, some compensating feature must be included in the design in order to achieve the desired degree of stability reliability.

Temperature is very important in circuits where transistors are used. The effective junction temperature of a transistor depends on the ambient temperature and the junction temperature rise due to the flow of current in the transistor junction. The characteristics of a transistor are such that its functions in a circuit will vary with temperature, and if such variations in the output of a particular transistor, for instance, would cause or contribute to a degradation of equipment performance to the degree that the equipment would be rendered unreliable, then a compensating design measure such as a feedback circuit to maintain suitable operation over a temperature range is necessary. The design engineer must recognize which components and circuits are critical, analyze the output of a critical component over temperature ranges, and design the circuit to compensate for the undesirable variations. The project engineer should be knowledgeable of the critical areas of a design and should be prepared to direct design action where necessary in order to achieve the required reliability objectives.

17.4 Selection of Components for Reliability

The specification and selection of components that will permit the realization of the reliability requirements of a subsystem is a second factor of reliability design and is probably the more important. Before the selection of a component can be made, the engineer must establish what are to be the criteria for the selection. Thus, the reliability analysis for application to the particular component under consideration must be interpreted and the individual component specification established. With the specification established, the component can be fabricated in-house or purchased from a supplier.

For the subtractor circuit of the Shadow Computer, a review of the reliability analysis indicates that a particular component must meet specific accuracy and stability requirements during continuous operation over a period of 24 hours. These short-term operational requirements can be achieved through the use of quality components as well as the design of special compensating networks which will function to correct any accuracy and stability deviations as the equipment is operated under conditions of different temperature ranges. In other words, practically any component, regardless of quality, can be expected not to fail if operated for the 24-hour test period. However, the design and reliability engineers must specify and select components that will not exceed a specific failure rate over the period identified as the normal equipment utilization period illustrated in Figure 17.1.

In examining any particular component of a design to establish what is to be its reliability specification, the design engineer must determine what type of component malfunction would render the equipment inoperative or result in a catastrophic breakdown.

The normal concept of failure is the obvious breakdown such as that resulting from the snapping of the drive shaft of an engine. For electrical equipment, failures usually occur in two models—either an open or a short. The design engineer must then determine for the component under analysis whether the occurrence of a short or an open or either type of component failure would result in a catastrophic failure of the equipment. If it is estimated from the specification of the component and the application in the circuit that the probability of component failure due to a short is 70 percent and that due to an open is 30 percent and if the function of the component will result in catastrophic equipment failure only when the component failure is a short, then the 70-percent figure would be used in the reliability design.

One facet of component reliability is the failure rate expressed as the rate per specific number of hours. Another method of expressing failure rate is the expected life expressed in hours under rated design conditions. The period of life of different components in a subsystem would be scaled to a common base for purposes of calculation. The expected life of a particular resistor in a circuit might be 20,000 hours. This can be expressed as a failure rate of 1 in 20,000 hours. A transistor having an expected life of 10,000 hours would then have a failure rate expressed as 2 in 20,000 hours.

In a calculation of the effective failure rate, the failure rate of a component which is based on design conditions of operation such as rated load and specific ambient temperatues must be modified by factors which reflect the operating conditions of the subsystem design that would utilize the component. In reliability engineering, these factors are referred to as stress conditions, which fall into two categories, namely, the maximum-limit and time-function conditions.

Failures as the result of maximum-limit conditions occur when the stress imposed on a component reaches the maximum limit of the component capability. The failures resulting from the time-function conditions occur more or less as the result of fatigue after a period of time.

For the electrical components being analyzed on the subtractor subsystem of the Shadow Computer, the maximum-limit conditions would generally be stresses such as voltage, current, vibration, and shock. The time-function conditions would involve temperature, current as it relates to heat-producing characteristics, fatigue due to vibration, and in a general way, the aging process. The failures due to time-function conditions will occur in decreasing periods of time as the level of the stress is increased.

The design and selection of a component which will meet the maximum-limit stress conditions is straightforward. This effort requires that the engineer calculate all possible sources of stresses that will safely withstand the worst possible condition. Basically, this involves the traditional engineering analysis and the application of a suitable safety factor. The same principle would apply whether a bridge or a radio receiver is being designed.

The time-function stresses involve a special analysis for reliability which must consider the magnitude of a stress, the failure mode that would result in a catastrophic failure, the meantime between breakdown, and other factors.

Table 17.2 illustrates how the analysis of various factors would be tabulated for the application of various components of the subtractor system design. A failure of resistor R1, for instance, would always be an open so that the probability of a short occurring would be zero, and an open would occur 100 percent of the time. The effect of an open for R1 would be a failure as indicated by F. Capacitor C3, if it failed, would experience a short 40 percent and an open 60 percent of the time. In other words, if a failure occurred for C3, the probability of a short occurring would be 4 out of 10 and of an opening occurring would be 6 out of 10. A failure of C3 as a short would render the subtractor system useless or be catastrophic (indicated as an F), whereas a failure of C3 as an open would be trivial and not affect the system (indicated as an N).

The information in Table 17.2 establishes the estimated statistics regarding the relative impact of failures of different components on the trainer reliability. The design and reliability engineers must now make an estimate of the probability of failure of each component, or, expressed a different way, the expected failure rate over a particular period of time.

A vast amount of data has to be collected relating to the results experienced with running life tests on various types of components, such as resistors, capacitors, and transistors. For instance, results of tests on transistor life operating at different temperatures are summarized in Table 17.3.

TABLE 17.2 Tabulation of Effects of Component Failures on Shadow Computer Subtractor Circuit

	Short		Open	
Component	Probability, %	Effect on system	Probability, %	Effect on system
Resistor R1	0	—	100	F
Resistor R5	0	—	100	F
Capacitor C3	40	F	60	N
Capacitor C7	80	F	20	F
Capacitor C8	70	N	30	F
Mode D6	50	F	50	F
Mode D7	50	N	50	F
Transistor Q6	50	N	50	F
Transistor Q7	60	F	40	F
Relay contact R2 ..	10	F	60	N
Relay contact R2 ..	10	F	20	F

TABLE 17.3 Summary of Typical Transistor Life Operating at Different Temperatures

Temperature	Life, hours
200°C	2
150°C	200
100°C	4,000

The stress factor or ratio relates to the percentage of rated load to which a component is subjected. For instance, a transistor delivering twice its rated load may have a life expectancy of one-tenth normal life. In like manner, a transistor operating at one-half its rated load may experience a life expectancy of five times its normal life. The stress ratio expressed as the operating load over the rated load is important in calculating the life expectancy or failure rate of a component. For application to the calculation of net failure rate, the stress factor is converted to a figure identified as the stress life factor. For the transistor in the above example, whereas the stress ratio is 0.50, indicating operation at one-half

TABLE 17.4 Tabulation of Factors for Reliability Prediction: Subtractor Circuit of Shadow Computer

Part	(1) Quantity	(2) Stress ratio	Stress life	(3) Temperature factor	(4) Failure rate per 10,000 hours	(5) Probability factor	(6) Net failure rate per 10,000 hours
Resistor R1	5	0.5	0.2	0.7	0.8	1.0	0.56
Resistor R2	1	0.7	0.3	0.6	0.7	1.0	0.13
Capacitor C3	2	1.0	1.0	0.8	0.9	0.4	0.58
Capacitor C7	1	0.6	0.3	0.7	1.2	1.0	0.14
Capacitor C8	4	0.5	0.2	0.7	1.2	0.3	0.19
Diode D6	1	0.3	0.1	0.6	1.9	1.0	0.11
Diode D7	3	1.0	1.0	0.6	2.2	0.5	1.98
Transistor Q6	2	0.7	0.4	0.5	2.7	0.5	0.64
Transistor Q7	4	0.5	0.3	0.7	2.7	1.0	0.23
Relay coil	1	0.8	0.7	0.8	0.6	0.7	0.19
Relay contact	1	0.8	0.7	0.8	1.0	0.3	0.22
							4.97

Total failure rate .... 4.97/10,000 hours
Mean time between failures.... 2,000 hours

rated load, the life would be increased by a factor of 5 or the failure rate reduced to one-fifth normal rate. Thus, the stress life factor would be 0.20.

Table 17.4 illustrates a tabulation of the main reliability factors and the simplified application of the various reliability factors to derive the net failure rate of each type of component. As indicated in the table, the net failure rate is derived as the product of the factors in columns 1 through 5. By calculating the sum of failure rates and dividing the time base by this sum, the meantime between failures of the subtractor circuit would be established.

The net failure rate of the complete system can be estimated by making a reliability prediction of each individual subsystem and deriving the failure rate or expected life between failures of the equipment itself as a statistical composite of the failure rates of each subsystem. If, for example, there were ten subsystems of a piece of equipment, each of which had a meantime between failures of 2,000 hours, then by the most elementary and crudest calculation, the meantime between failures would be 2,000/10 hours or 200 hours. If the equipment was comprised of twenty subsystems, each with a meantime between failures of 2,000 hours, the reliability figure of the equipment would be 100 hours. The general conclusion is that as the complexity of an equipment goes up, the reliability factor goes down. All other factors are kept the same. Another conclusion that can be reached is that if a high reliability is desired in a complex piece of equipment, some of the reliability factors listed in Table 17.4 can be improved. For instance, the stress ratio can be lowered by operating a component at a lower percentage of its rated capacity, which would result in a longer expected life for the component.

17.5 Reliability Review

Even though reliability engineers may participate in the design effort, their official function is advisory in nature. After design engineers complete the work on a particular subsystem, it is reviewed and approved by the project design engineers and is transmitted to the project quality coordinators for review by reliability engineers in their department. Reliability engineers would, in effect, be reviewing a design of which they should already have a detailed knowledge. Their review would verify that the design engineers have interpreted and included their recommendation in the subsystem design. Also, they would review the design to determine whether there is any area which was inadvertently omitted in the reliability analysis.

The review by reliability engineers and their staffs will cover three general areas of effort: component reliability, system analysis, and environmental tests.

The component reliability review involves a recheck on the specification terms, the thermal, loading, and other operating requirements of each component, and the investigation and verification of the information compiled on the Reliability Prediction Chart. The effort would, by necessity, consist primarily of a selective reliability spot check and analysis. For example, on the reliability prediction tabulation in Table 17.4, the reliability engineer would be more interested in verifying the stress ratio of resistor R1 of which there are five units than of R2 of which there is only one unit. The reliability check would involve essentially a repetition of the analysis made by the design engineer. For example, arriving at the conclusion that for a stress ratio of 0.5 for resistor R1, the stress life should be 0.4 instead of 0.2 (indicating a failure rate twice that which was derived by the design engineer), the reliability engineer would not approve the design in the particular area and it would be returned for resolution by the engineer.

In like manner, the reliability engineer would analyze and reexamine the approach and circuit details submitted by the design engineer to verify whether adequate compensating circuits have been included to guarantee stability and accuracy of the equipment.

The achievement of reliability for any piece of equipment is a continuing task, even after the equipment is in operation in the field. Reports of all failures or troubles should be referred to the reliability group for analysis and correlation. If such reports indicate a recurrence of a particular type of difficulty, a design analysis should be made and a modification to the equipment should be implemented to correct the difficulty and thereby achieve the desired improvement in reliability.

17.6 Maintainability Principles

As noted in Section 17.1, the maintainability of a piece of equipment is a measure of its features which permit expeditious repair when failure occurs. The quantitative measure of maintainability is usually referred to as MTR. The term is usually further defined to establish what is to be considered a failure in the equipment, e.g., is the component that failed of the type that is carried as a spare part for the equipment?

The design for maintainability requires that basic features such as access to the inner components be provided, test points be identified, instrumentation and tools be available, assembly tester and power and air pressure supplies be provided, adequate spares be stocked, etc. In other words, the equipment design and facilities must be so designed and provided as to provide optimum capability to make repairs and get the equipment to function normally after a failure occurs.

A significant parameter which establishes the measure to which the equipment is capable of being utilized is the availability factor. Availability is expressed as a decimal and is established by the values of MTBF and MTR. The availability factor is determined by the following relation:

$$\text{Availability A} = \frac{\text{MTBF}}{\text{MTBF} + \text{MTR}}$$

Expressed in another way,

$$\text{Availability} = \frac{\text{Operating Period}}{\text{Operating Period} + \text{Repair Downtime}}$$

For the RLS the specified requirements for MTBF and MTR are noted to be as follows:
   MTBF (Mean Time Between Failures) = 100 hours
   MTR (MEAN Time to Repair) = 1 hour
   Therefore the availability factor of the trainer would result in a value of

$$A = \frac{100}{100 + 1} = 0.99$$

The effectiveness factor is a measure of the probability that the equipment will be ready and capable of performing its function and will not experience failure during its mission period.

It is determined by the value of the equipment reliability, the value of the availability, and indirectly the ability to effect repairs.

For a predetermined level of performance, effectiveness can be expressed as the product of reliability and availability.

$$\text{Effectiveness} = R \times A$$

Because of the qualifications related to a specific mission as noted above, the above expression for effectiveness should be used only when the conditions are preestablished. Effectiveness is generally used for military systems and equipment where equipment performance at a particular time is of vital and critical concern.

Figure 17.2 illustrates a typical family of effectiveness curves for a piece of equipment that might be designed for different values of reliability and availability.

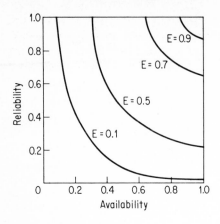

FIG. 17.2   Effectiveness curves for values of reliability and availability.

17.7   Summary

The general definition of reliability is the ability of the equipment to operate under specified conditions for a specified period of time without interruption due to equipment failure or degradation of performance. The specific definition of the length of time that the equipment is to operate and what constitutes equipment failure are dependent upon the type, function, and design of the equipment to which the reliability is applicable.

The reliability of any equipment is based on the ability of individual components to perform as required and on the ability of the design of the various critical subsystems to compensate for changes in load, temperature, and other conditions that vary during equipment operation. The degree of reliability will differ, depending on what period in the equipment life reliability is observed. The three basic periods are infancy, normal utilization span, and obsolescence. Due to different factors, a piece of equipment would experience relatively poor reliability during the infancy and obsolescent periods.

The basic measuring tool of reliability is the meantime between failures. The two modes of reliability failure are catastrophic failures causing the shutdown of a system and degradative failures in which the equipment suffers due to progressive malfunctions causing a loss of accuracy or stability.

The implementation of reliability is accomplished by a special reliability and quality-control group which, in the illustrated case, is identified as the project quality coordinator. A reliability engineer would be generally assigned to advise the engineer designing a particular subsystem regarding the reliability requirements of the particular area. The reliability requirements for the system are identified in the reliability analysis. Each system is analyzed as to which components are critical to the operation of a system, what type of failure of the component would affect the system operation, and what the probability is of the particular failure occuring for the component.

A further analysis of each component is made in a reliability prediction study. The reliability prediction study considers such factors as the quantity of each component used, the stress ratio, the temperature factor, the failure rate per period of time, and the probability factor mentioned above. As a result of the reliability prediction analysis, the estimated meantime between failures of the system can be derived by a rough statistical operation.

Maintainability is a measure of the ability to repair a piece of equipment when failure occurs.

Availability is a measure of the utilization capability of a piece of equipment and is established by the equipment reliability and maintainability.

For a specified set of conditions, the factor of effectiveness, which is the product of reliability and availability, can be established.

PROBLEM

A particular subsystem of a complex computer is comprised of the following components:

4 Resistors R1	6 Diodes D1
3 Resistors R2	2 Diodes D2
2 Capacitors C1	5 Transistors Q1
4 Capacitors C2	2 Transistors Q2

It is assumed that failures to the resistors will always occur as shorts. Failures to capacitors will occur as shorts 40 percent of the time and as opens 60 percent of the time. Failures to diodes and transistors will occur as shorts and opens equally.

Catastrophic failures will result when any of the following occurs: resistors short, capacitors open, diodes short, transistors open. All components are operated at 50 percent of their rated characteristics. It is assumed that the relation between stress or temperature and component life is directly proportional. The failure rate per 20,000 hours for each category of components is as follows:

Resistors	1.5
Diodes	1.0
Capacitors	1.2
Transistors	0.9

Develop the component failure and reliability prediction tables, and compute the meantime between failures for the subsystem described above.

Chapter Eighteen

Production and Quality Control

18.1 Production of Prototype Equipment

Production as a subject covers a very broad area and involves a multitude of concepts and disciplines. The assembly-line production of a toaster is vastly different in its control and organization from the production of a custom-built reactor for a power plant. Since the production of a prototype device necessitates a close liaison with engineering, project engineers would have to implement a line of communication between the engineering and production departments with themselves as the link between the two departments. It is not to be inferred, however, that project engineers or engineering departments would concern themselves with the many routine and standard facets of production. The engineering department would be primarily concerned that the production methods do not adversely affect the fidelity of the system or component operation as designed. It would also render consultation regarding production problems that arise which might affect the equipment design. For instance, a soldering process might expose a particularly delicate component to excessive heat, thereby degrading its function. The design engineer would be called upon to substitute the component or work out some procedure with the production department whereby the production difficulty could be overcome.

The project engineer, while concerned with problems affecting the equipment as far as its ability to function as designed, is also interested in the status of the production schedule and the accumulation of production costs. Thus, the reporting procedures and control techniques would be implemented to guarantee to the maximum degree the objectives of the contract schedule, cost allocation, and technical requirements of the equipment.

18.2 Production Planning and Control

The project engineer would delegate the responsibility and authority to the project production coordinator to implement the necessary planning and control for the production of the equipment that is required.

The planning effort involves the coordination and scheduling of equipment, personnel, and materials required to accomplish the task as scheduled. The control effort involves the setting of the plans in motion by releasing the orders and monitoring, inspecting, and recording the progress so that a continuous comparison between the planned and actual results can be made. Since project engineers are responsible for seeing that the manufacture of the equipment is accomplished on schedule, is within the budget that has been set, and meets the specification requirements, they would work closely with project production coordinators when the plans are generated. They must also ascertain that the methods and means of control are established so that the production effort will result in meeting the various goals.

The manufacture of the first article or prototype piece of equipment which is the result of a development and engineering effort involves special planning. One unique feature is that all the detailed information and drawings generally will not be available at one specific time. The program schedule, which is usually critical, often requires that manufacture be accomplished in phases as the required information and drawings are released for production.

Because of the piecemeal nature of the manufacture of prototype equipment, the planning and control must be supported by some logical and feasible division of the equipment elements. The division of the production effort into elements would be analogous to the work package concept of the PERT plan described in Chapter 11. The work package generally relates to the circuits and elements contained in a chassis (when dealing with electronic equipment). However, with the use of integrated circuits and other high compact modules, a single chassis can accommodate several different systems—each of which is described in a separate work package. The design of the landmass simulator, for instance, would combine the Shadow Generator and the directivity effects on a single chassis, and the production plans must provide for such a combination to be performed as a single task.

Project production coordinators must establish a phasing chart for the different areas of work to be accomplished under their cognizance. The start of the production in each area is contingent upon receipt of drawings from the drafting section of the engineering department which, in turn, is contingent upon the completion date of the design effort. There are certain exceptions to this procedure which relate to the production of different standard components and modules which the project engineer and the project design engineer have established; these would be used in the equipment even before the design is completed. Such components include connectors, sheet metal work for consoles, standard servo units, and similar categories of items. These items, upon completion, would be stored as components to be used at a future point in the program.

The creation of a phasing schedule is a coordinated effort among the various departments that are involved. In this case, the individuals involved are the systems engineer, project design engineer, project production coordinator, and the schedule and cost coordinator, all of whose contributions and inputs are coordinated by the project engineer for the final schedule.

One point of interest that should be noted relates to the advantage of having the drafting section under the cognizance of the project design engineer. Any slippage in the design effort and/or drafting effort is the responsibility of one individual who can apply whatever accelerated effort may be required to get the release from drafting on time and thereby avoid delays in starting the production cycle.

As far as the production effort is concerned, the initial concern of the project engineer is that the drawings in the various areas are completed and released to the project production coordinator as scheduled. In order to effect adequate preparation, the project engineer would receive periodic Schedule Prediction Reports from the schedule and cost coordinator that are derived from information compiled by the project design engineer. The reports would give information regarding each of the areas of the phasing chart. The prediction reports similar to the example shown in Figure 11.6 would alert the project engineer of any difficulties that exist and permit possible remedial action as early as possible.

It should be noted that any information regarding the schedule and cost of the project would be received and correlated by the schedule and cost coordinator prior to being forwarded to the project engineer. There are many advantages to such a procedure, especially if the company has more than one program under contract. Schedule and cost coordinators, having the overall picture of the operation of the company, can present an evaluated and objective report of the status of the program. They are also able to relieve the various department heads and the project engineer of the details of correlating the information into the necessary form.

Once the production effort is initiated in any area, the project engineer is primarily interested in whether the schedule is being maintained, whether the accrued costs are within the budget, what courses of action are available if either the budget or schedule is not being met, and what effects on the cost and delivery any possible design change might make.

Project engineers would derive their basic information from three reports similar to those discussed in Chapter 11, namely, the Schedule Prediction Report (Figure 11.6), the Cost Prediction Report (Figure 11.5), and the Line of Balance Reports (Figure 11.7 and Figure 11.8). The information for the reports would be accumulated by the project production coordinator by means of the cost-accounting system that has been adopted for the project and would be correlated by the schedule and cost coordinator for presentation to the project engineer.

In establishing the schedule status at any particular point in time of the program, consideration must be given to the characteristic schedule curve of accomplishment. This graphic presentation is referred to as the S curve.

Figure 18.1 illustrates the S curves for the production and assembly schedule of the optics system and the detector system of the RLS. The solid lines are the planned schedule for each system. The dotted lines represent the actual progress of the production of each system.

Reference to the Figure 18.1 will reveal that as of December 1, the detector system is 85 percent completed, whereas it should be 95 percent complete. The optics system on the other hand is ahead of schedule since

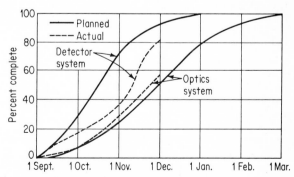

FIG. 18.1 Characteristic schedule curve of accomplishment (S curves) for detector and optics system.

as of December 1, it is 60 percent complete as opposed to about 50 percent, as planned.

One important characteristic of the S curve that should be noted is of course its shape. At the start of the work, the slope is almost horizontal and increases rapidly to almost a vertical, and decreases to a horizontal again. The shape reflects the fact that at the start of the work, relatively little is accomplished because of necessary planning. During the middle portion of the period, progress is made most rapidly and then slopes off again to the finishing up of the effort.

The slope of the optics system production S curve indicates that the

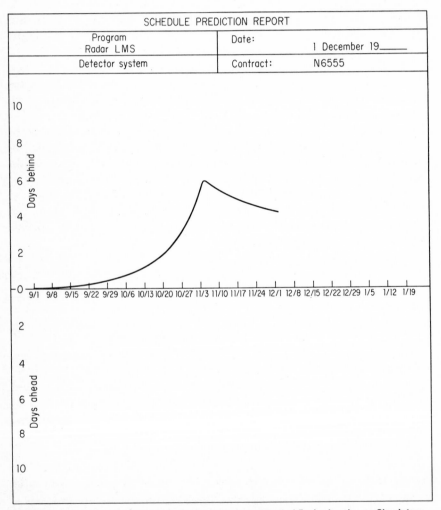

FIG. 18.2   Prediction report for production of detector system of Radar Landmass Simulator.

nature of the work requires a large amount of planning and preparation so that its slope is relatively shallow during the initial phase of the effort as compared to the detector system S curve.

The prediction reports would be derived from the S curves. Figure 18.2 illlustrates the production prediction schedule report for the detector system. In a comparison of Figures 18.1 and 18.2 the correlation of the slippage in schedule can be noted, and the indication of the results of the corrective action that had been taken is shown by the reversal of the slope of the curve in Figure 18.2. It should also be noted that the corrective action had not been taken soon enough, nor is it adequate to permit the scheduled completion date of January 1 to be met.

The project engineer would use the information presented in the forms as discussed to report to higher management and take other more drastic corrective action as necessary.

The same type of reporting media is used for the comparison of actual costs with budgeted figures. The Cost Prediction Report is prepared for the project engineer so that the trend of production costs in various areas can be observed and so that the necessary corrective actions may be taken if the cost experiences are unfavorable. The possible courses of corrective action that would be taken in the event that the cost or schedule picture is unfavorable or shows indication of becoming unfavorable would be a function of the analysis of the situation and the conditions that prevail. The main point that the project engineer should recognize is that when there are indications that there will be a slippage in the schedule or that the cost will exceed the budgeted costs, corrective action is required as early as possible.

18.3 Quality Control

On a project similar to the RLS, quality control in production embraces two general areas. The areas in question are the quality of the module assembly and the quality of various components.

The objectives of reliability excellence are synonymous in many respects to quality so that the achievement of reliability often will automatically result in the achievement of quality. The excellence of the final product will depend to a large degree on how well reliability is incorporated in the design and how well quality is achieved in the manufacturing and assembly process.

The factors that constitute quality of a product will differ for different types of equipment. For instance, an electronic system operating at low frequencies can be built to meet satisfactory quality standards with little or no attention given to shielding, length of leads, etc. However, for a high-frequency system, these factors are very important, and failure to exercise the proper control in the production effort in these areas will result in a product that, as far as quality is concerned, is unacceptable. The quality-control engineer must therefore exercise proper controls to guarantee that the production department recognizes all problems and that it achieves the quality objectives that have been specified. The project engineer is particularly concerned that the project production coordinator implement a program that will achieve the required quality objectives.

Another area of quality that concerns the project engineer is component quality and the means by which it is achieved. Quality control, in the traditional sense, involves the statistical sampling of a component as it is produced or delivered from a supplier.

When a volume of a component is produced or procured, it is often impractical to check every unit to determine whether the components meet the specification requirements. Therefore, a calculated risk must be assumed by the user as to whether a particular batch meets the specification requirements.

The normal distribution curve discussed in Figure 11.3 would be used as the basis for the statistical sampling of a production run of a component. The tolerance that production coordinators are willing to accept and the risk they are willing to assume would be correlated with the standard deviations of the curve to achieve the sampling criteria that are desired.

18.4 Quality Control and Management

There are four general areas of quality control which are of concern to the project engineer.

One is to certify that the machinery used for production and the tools and instruments used for measurement and inspection have the necessary quality and accuracies.

The second is to establish adequate provision for complete and accurate instructions, drawings, etc., with which to do the task. In order that the required quality control be carried out, the responsible workers must have the proper information.

The third is an adequate plan of inspection. This includes not only the inspection of in-house shop production but proper inspection of purchased components. In order to implement the inspection process, the proper statistical methods must be used as mentioned earlier.

The final area is to maintain a consistent standard of implementing quality control. In particular, this means that a periodic checking and calibration of the instruments and tools used for inspection is necessary so as not to permit a deterioration of the quality-control program.

18.5 Summary

The production plans and techniques will vary for different types of equipment. For a development prototype piece of equipment, the production plan would be in essence a custom-built device.

Project engineers would be primarily interested in the cost and schedule status of the production effort and would implement a reporting system by which such information can be transmitted to them.

Project production coordinators would derive a phasing schedule for the various subsystems of the equipment. The main production effort for each subsystem would be initiated upon release of the manufacturing drawings from the drafting department.The three areas of concern of the project production coordinator are that the cost is kept within the budget figures, that the production schedule is maintained, and that the quality standards are achieved.

The cost and schedule status of the program is controlled by frequent periodic analyses of the actual program status with the planned program. The quality of the equipment is controlled by implementing the appropriate quality-control program for the production effort.

Chapter Nineteen

Test and Checkout

19.1 Test Objectives

The creation of any product requires that some procedure be adopted and followed for establishing whether the product is consistent with what was intended. For nondynamic items such as shoes or food products, the tests would represent quality-control checks relating to tolerances, grade of materials or ingredients, workmanship, etc. For a piece of equipment which is designed to perform in a specified manner, the tests would cover a much larger area and necessitate the measurement of dynamic results. Since the project engineer is generally involved with dynamic equipment, this text will discuss the type of tests that are necessary to establish whether the equipment being produced satisfies the requirements specified by the contract.

The test and checkout of equipment being produced as a development or a new design is a continuing effort throughout the program. The testing falls into two basic categories. The first relates to the verification tests which the contractor performs to be sure that the equipment will function as specified. The second category represents the acceptance tests which are performed by the customer and which determine whether the equipment meets the requirements of the contract and specifications.

The types of tests performed by the contractor would include the following:

1. *Breadboards:* To verify the merits of a particular circuit or module design

2. *Subsystem:* To establish the performance and compatibility of specific subsystems of the equipment

3. *Computer high level programming:* For verification of math model

4. *Integrated:* For verification of overall equipment performance

5. *Reliability:* To determine ability of elements, modules, and circuits to perform within acceptable reliability limits

6. *Quality and assurance:* To verify acceptability of components, workmanship, arrangements, etc.

The customer's primary concern is to determine whether the equipment specified in the contract satisfies the contract requirements. The type of tests that would be performed by the customer would include the following:

1. *Quality and assurance:* As above

2. *Equipment performance:* To establish the acceptability of the equipment performing as a completely integrated system

3. *Computer performance:* To determine if the memory, accuracy, and speed of the computer, as programmed, satisfies the specification

4. *Reliability:* To establish whether the quantitative values of reliability for the equipment can be met

5. *Maintainability:* To establish whether the maintainability criteria for the equipment can be met

6. *Environmental:* To establish whether the equipment satisfies the environmental requirements of the contract

19.2 Test Criteria

For a piece of equipment that is being engineered in response to a performance specification, the detailed test criteria must generally evolve with the design effort. The performance specification may provide equipment objectives such as general accuracy criteria, response times, etc., but the detailed characteristic performance curves, varieties of functions, and many combinations of performance modes are not generally expressed in the performance specification. Because of the nature of such projects the test criteria must be developed during the design phase and represent a mutually agreed upon vehicle for use in establishing the acceptability of the equipment.

Figure 19.1 indicates the sequence in which design data and test criteria are developed as a system is engineered. The performance specification identifies the objectives required from the equipment being procured and also identifies the source of data that is to be used in designing the equipment. Generally, the detailed data supplements the performance specification and the acquisition and use of the data is the responsibility that the contractor had assumed in the contract.

For the landmass simulation, for instance, the detailed data relating to the radar antenna beam pattern must be procured by Acmen Electronics in order to simulate the radar returns as required by the specification.

The design model is analogous to the system design discussed in Chapter 15. If a digital computer is used in the system, the design model would include the math model from which the computer program would be derived.

The next step in the evolution of the design and test criteria is the creation of the actual hardware design and the software or computer program. The verification of critical or problem design areas would be accomplished by the laboratory breadboard technique. Any corrective actions that might be found necessary from the breadboard results would be fed back to the design model and the process repeated.

When the digital computer is applied, a high-level computer language such as Fortran is used to verify whether the results intended by the math model are consistent with the performance specification and the detailed data that were used. Again, corrections to the math model and the computer program are accomplished by the feedback loop shown in Figure 19.1.

Once the design and computer programs (where computers are applied) are perfected, the equipment is fabricated, assembled, and ready for subsystem and system test. The test criteria should now be valid for establishing the acceptability of the equipment because they not only were derived from the design model that was the basis for the equipment design (and computer program) but were the corrections resulting from the breadboards and high-level computer programs that were fed into the model. Thus, if the development of the design and test criteria was accomplished systematically and faithfully, both the customer and contractor would not hesitate to accept the criteria for testing the equipment produced.

19.3 Types of Tests

The various types of tests that would be made on newly engineered or development projects that were mentioned in Section 19.2 are further described as follows:

Breadboards

A breadboard comprises elements of a circuit that has been assembled and instrumented at a test bench to verify whether the intended performance of the circuit will be met. Breadboards are generally used for electronic and electrical designs during the engineering phase.

Subsystem Tests

The subsystem tests are scheduled immediately upon release from the production department. The initial phase of these tests is the static tests in which power is applied to the equipment in gradual steps primarily to uncover any wiring errors. When the static tests reveal a subsystem that is sound, measurements for accuracy, stability, and other static characteristics are verified.

Upon completion of the static tests, the equipment is subjected to the dynamic tests which will verify response times, phase shifts, accuracies, and other characteristics of the subsystem. It is during the subsystem verification tests that necessary design fixes are incorporated. Since the subsystem tests are conducted in open areas, any equipment changes can be made with a minimum of effort.

The criteria for a subsystem test would be unique for the equipment being tested, and in many cases a degree of ingenuity must be exhibited in order to establish the identity of the test criteria. In the case of the landmass simulator, the establishment of the test criteria for the FSS as a subsystem might be considered. There are several parameters of the FSS which must be verified, such as the degree of stability of the light intensity and the accuracy of the bearing of the spot sweep. It would be up to the project design engineer to establish the method and the instrumentation that could be used to verify whether the equipment meets the requirements of the specification.

Computer Program Tests

Chapter 16, Digital Computer Applications, addressed the subject of computers in the broad terms of the computer speed, memory, and accuracy characteristics as it would ultimately be expected to perform. The earlier sections of this chapter discussed computer languages and specifically Fortran as one of several types of high-level languages used for computers.

The three basic types of computer languages are: machine language, assembly language, and high-level language, such as Fortran.

A digital computer must be told in the minutest detail what to do. Machine language is also referred to as absolute code and determines the step-by-step operation of the computer. Each step can be performed in less than a microsecond. The machine language program is unique to each category of computer and computer programmers working in machine language must be intimately familiar with the computer with which they are involved.

The assembly language is more universal than the machine language.

The programmer need not be intimately familiar with the particular computer and does not need to be concerned with the address of the instructions, etc. The assembly language is converted to machine language by a special assembly program. Most computer programs are accomplished in assembly language.

One of the most universal computer languages used today is the high-level language referred to as compiler language, such as Fortran. The computer program can be written in general mathematical expressions. The compiler translates the written language into machine language.

Because of the ease of writing a program in high-level language, the procedure is used in verifying the math model and generating the test criteria as discussed with reference to Figure 19.1. One of the reasons that high-level language is restricted to test phases and is not more universally applied as an operational program is that its use imposes greater demands on the computer memory and time. Therefore its use results in an inefficient application of a digital computer.

The verification of whether the computer memory and word length is adequate for satisfying the specification requirements can be accomplished by an analysis as discussed in Chapter 16. The time requirement can also be verified by an analysis, and the testing of computer operation can be accomplished by exercising the computer through the worst case sequence of timing operations that might be anticipated.

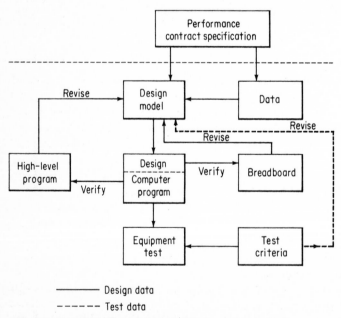

FIG. 19.1 Flowchart for development of design and test criteria.

Integrated Test and Checkout

When all subsystems of the equipment have successfully passed the subsystem verification tests and have been interconnected to form the final equipment configuration, the integrated test and checkout program starts. The criteria to be followed was discussed above. It is this criteria which, by mutual agreement between the contractor and customer, determines whether the equipment is acceptable or unacceptable.

For an equipment composed of a complex electronic and optical system, the test procedures and criteria could, in many areas, be rather involved. As an illustration of the complexity of some of the test procedures, the test that might be used to verify the accuracy of the Shadow Computer is presented as follows:

SHADOW COMPUTER VERIFICATION

1. PROCEDURE: Align the transparency so that the test step wedge would be illuminated. Scan the step wedge from the high to low altitude representation.

2. Set the radar control switch to PPI and the range at 75 miles.

3. Disconnect wires labeled 5A2-B6-J6-50 and 3B6-D2-51-42.

4. Set the aircraft altitude at 5,000 feet.

5. Observe readings of terrain elevation signal $e(T)$ and aircraft altitude signal $e(A)$ on oscilloscope from readings derived from points 2A4-C3-A-1-J1 and 1A2-B3-B6-J5. The shadow blanking plus is +9 volts.

6. Complete the readings of Figure 15.1 taking measurements based on the following parameters:

$a.$
$$e(t) \text{ feet} = \frac{e(t) \text{ volts}}{50 \text{ volts}} \times 50,000 \text{ feet}$$

$b.$ $e(A) - e(t)$ = altitude of aircraft minus elevation of terrain at start of shadow

$c.$ $\Delta e(t)$ = terrain elevation different at start and end of shadow

$d.$ T = time from sweep start to shadow start in microseconds

$e.$ ΔT = time from shadow start to shadow end

$$T = \frac{e(t)}{e(A) - e(t)} \times T$$

$f.$ Error of ΔT

$g.$ Required accuracy—5%

The above example indicates that the integrated test and checkout procedure is much more than making visual inspections or taking superficial readings of measurements. A thorough and valid test procedure often involves using complex testing equipment, using a procedure of effecting temporary modifications to the equipment to gain access to the test points, taking the readings by operating the test equipment, and

TABLE 19.1   Test Data and Results of Shadow Computer Verification

$e(A)-$ $e(t)$	$\Delta e(t)$	T Micro- seconds	$\Delta T-$ Calcu- lated	T Measured	Error	Accuracy
10,000	2,000	40.5	7.6	39.0	1.5	3.7%
15,000	1,500	30	5.2	29.0	1.0	3.4%

making the necessary calculations to determine whether the readings obtained from the equipment fall with the tolerances of the equipment as illustrated in Table 19.1.

From Table 19.1, it can be seen that the accuracies of the equipment as the result of two readings are 3.7 percent and 3.4 percent, which are within the 5 percent tolerance noted.

Because of the critical nature of the test phase, project engineers must monitor the program very closely. Usually, they are present on the test floor and conduct daily status conferences to determine difficulties that were encountered and make decisions on problems of a major nature that have been referred to them.

The test effort would be initiated upon the issuance of the Test Engineering Order, which would include the test criteria documents, schedule, and other necessary information, to the group leader of the testing team.

If the testing program reveals any design deficiencies, corrective action by an engineering change order would be instituted and the revised design information reflected in the test criteria documents as required.

Reliability Tests

The test for reliability is the most time-consuming and is generally impractical during the contractor test and checkout phase of the equipment. A review of the engineering design analysis that was used by the contractor for providing equipment that would meet the reliability requirements as discussed in Chapter 17 will provide a good statistical basis for establishing the reliability feature of the equipment. Customers, during their checkout, will often maintain a record of the failures experienced by the equipment in conjunction with the integrated test program. Such a procedure establishes when the reliability requirements of the equipment are acceptable.

Environmental Tests

The environmental tests usually require the use of special test facilities which can be used to subject the equipment to the various tests. Typical environmental tests for electronic and similar equipment are vibration, humidity, and radiation tests. The test procedures and methods are fairly standard and are performed by specialized laboratory technicians. The primary concern of the project engineer is how well the equipment meets the tests and if any discrepancy occurs what corrective action is necessary.

The reliability engineer would play a very active role in observing the environmental tests, since these tests would provide vital evidence as to how well the reliability requirements of the equipment have been met.

Upon completion of the environmental tests, the results of the entire test program are presented to the customer for review and approval. The customer will usually repeat selected tests of the criteria especially in critical or questionable areas to substantiate that the equipment is acceptable. Upon formal acceptance, the equipment is prepared for shipment and the basic functions of the project engineer are concluded.

19.4   Scheduling of Test and Checkout

The test and checkout of the finished product is, in many respects, the most critical effort especially from the point of view of the schedule. Because the integrated system test and checkout always occurs at the end of the program, any unforeseen difficulties that occur have very little, if any, slack period to be corrected. A further impediment to swift corrective action is the fact that the test and checkout occurs when the system is packaged in its final configuration. As a result, the physical space to troubleshoot, take corrective measures, or incorporate modifications constitutes a difficult and time-consuming task. When difficulties of even a relatively minor nature occur during the testing and checkout of the integrated system, the project invariably slips to the point where delivery of the equipment in accordance with the schedule is impossible.

It is vitally important that detailed plans are made for the test and checkout effort. For a prototype system which involves development, the engineering department would perform the tests. The schedule and plans for such a test program would be made by the project design engineer, the project engineer, and the schedule coordinator, with contributions by the various engineering group leaders who would be doing the actual work.

19.5   Summary

The testing and checkout function that is necessary for the equipment being produced on a project is a continuing function throughout the cycle of design fabrication and performance verification. As the design is created and corrected, the test criteria for the product would be developed. The end result is that there must be a consistency among the equipment performance, the data and requirements upon which the design is based, and the criteria for the test and checkout.

The various types of tests that are required include breadboarding, subsystem tests, computer program verification, integrated tests, reliability tests, and environmental tests. Those tests performed by contractors for their own knowledge are referred to as verification tests. Those performed by the customer are the acceptance tests.

The test program of modern prototype equipment requires that the engineering department perform the tests since personnel involved in the equipment design are the only ones who are in a position to implement correction of deficiencies that the tests may uncover.

PROBLEM

Plan a test and checkout program for the pumping unit described in the problem for Chapter 15. Describe the following general types of tests to be performed on all subsystems: subsystem static tests, subsystem dynamic tests, and integrated system tests.

Chapter Twenty

Supporting and Monitoring Items

20.1 Identity

For any procurement of complex equipment, the contract will require the delivery of various support and contract monitoring items. The supporting items are identified with and necessary to the installation, maintenance, modification, and operation of the major items under procurement. Examples of categories of such items are manuals for the equipment, spare parts, engineering drawings, special tools and test equipment, and courses of instruction for the customer's maintenance and operating personnel.

In addition to the support items, the contract may require delivery of specific engineering and status reports, technical liaison services, and similar items deemed necessary by the customer in order to effectively monitor the technical progress, schedule, and, where applicable, the cost status.

The various support and monitoring items are usually referred to as contract side items and encompass all deliverable contract items with the exception of the main item of equipment.

The side items are all too often given only minimal attention, with the result that the contractor, in underestimating the effort required to deliver the side items, can suffer financial losses on a contract, even though the main contract item may have been successfully completed and delivered. The difficulties noted above should not be surprising when one considers the cost of providing the side items can often exceed 15 or 20 percent of the total contract selling price.

The side items which are usually the cause of difficulties are the engineering design reports, installation and maintenance manuals, and drawings.

The specification and contract requirements governing the format and content of the reports and manuals are usually very explicit and detailed. Contractors who have experienced losses and/or damaged performance reputations due to these side items can generally point to the following as reasons for their unfortunate experiences:

1. Failure of project engineers or their staffs to study or comprehend adequately the detailed requirements of the side items.

2. Failure of project engineers to assign qualified personnel to producing each of the side items or to subcontract those side items if the contractor neither has the facilities nor personnel for the completion of the item.

3. Underestimate of the scope or complexity of the task by the project engineer. Because of this, the project engineer assigns such tasks as a collateral duty to existing personnel who may not be qualified or motivated to complete the task in a successful manner.

The contract schedule for the landmass simulator describes three side items to be delivered in support of the simulator which are as follows:

Item 2a: Terrain Readout System Engineering Report
Item 2b: Signal-processing System Engineering Report
Item 2c: Display-system Engineering Report
Item 3: Engineering Drawings
Item 4: Installation and Maintenance Manuals

In addition to the side items required by the contract for the landmass simulator, contracts for other types of equipment will require delivery of special side items necessary for the equipment support.

20.2 Engineering Reports

The engineering reports should logically be the product of the design engineers who are responsible for the design effort in each pertinent area. However, the actual task of writing the report might be more expeditiously performed by technical writers working in close liaison with design engineers. Technical writers would normally be associated with a special technical writer group responsible to the project design engineers from whom they would receive their writing assignment, schedule, and directions.

Prior to initiating their work in planning a particular engineering report, technical writers must study the requirements of the specifications and contract schedule to establish what areas of the report must be emphasized, what degree of detail the report should use to describe the subject matter, what format and arrangement of information should be used, to what extent sketches, block diagrams, and sample calculations must be used, and other similar information.

In addition to the content of the report, technical writers must take cognizance of the contract submission date of the report and the number of hours that have been allocated in the program to complete the document.

Upon receipt of their assignment for a task, technical writers must study the proposal used to obtain the award of contract and find out whether any changes in design approach or content have been adopted since the submission of the proposal. After doing the preliminary study, technical writers follow the organizational procedures for drafting an outline and use the information provided by the design engineer for writing the engineering report.

After the completion of the final draft, the report is transmitted to the design engineer, project design engineer, and project engineer for approval prior to formal printing and reproduction.

Since the approval of an engineering report generally constitutes a design freeze for the equipment in the particular area and since the entire program schedule is based to a large extent on establishing the design freeze by a certain date, it is essential that everything possible be done to obtain the report approval by the program deadline date for design freeze.

The design freeze establishes what is to be the technical approach of a particular area, and the amount of detail in the engineering report should be consistent with what constitutes the tier II or III detail, depending on the specification and contract requirements.

For the RLS contract, item 2a, for instance, requires that the Terrain Readout System Report be submitted within 20 weeks from the date of contract, and the review time is specified as 30 days or 1 month. If the program schedule of Acmen Electronics reflects a design freeze date for the terrain readout system at 6 months after date of contract, then approval of the report must be obtained on the first submission. A rejection would require resubmission, and if the reason for the rejection is based on the disapproval of the basic design approach or some significant facet of the design, the contractor could not risk proceeding on some unapproved design approach. The program schedule must therefore be adjusted to reflect a later design freeze date for the terrain readout system. The postponed freeze date would act to delay the starting dates of all subsequent phases.

Usually, the rejection of engineering reports is due to insufficient design detail, omissions, failure to follow format, or some other reason of a noncritical nature. In such cases, the design freeze date can be maintained (at some risk to the contractor), but costly and time-consuming effort must go into correcting and revising a report which could quite as easily have been correct on the first submission if adequate preparation had been made.

20.3 Manuals

The preparation, writing, and publishing of manuals for delivery as an item of the contract probably represents the largest single side item of the contract. For a complex piece of equipment, the maintenance manual represents a large volume. The amount of detail, schematics, etc., will be dictated by the applicable specification on the format, degree of detail, and other requirement. In general, the maintenance manual must contain the degree of detail and supplementary documents such as sketches, etc., which will permit average maintenance personnel, knowledgeable in the technical area represented by the equipment, to be able to troubleshoot, repair, and maintain the equipment by referring to the manual alone.

The writing and printing of a maintenance manual is a specialized task, and except for the very large organization that may have its own personnel and facilities for the task, maintenance manuals are subcontracted to publication firms.

Regardless of whether the task for producing maintenance manuals is accomplished by in-house effort or subcontracted, the design engineers must provide the required technical information to the technical writers and maintain close liaison with the writers.

Those undertaking the writing of the manual must be provided with the specification requirements, the schedule for submission of the different drafts, and all other pertinent information. Because of the scope of the task and the amount of effort required, those undertaking the task of furnishing the maintenance manuals must carefully study the specification and contract requirements and make all possible preparation to achieve approval on the first submission.

In the case of the RLS, the Acmen Electronics Company would subcontract the task of furnishing maintenance manuals. Because the simulator contains several relatively unrelated technical areas such as the optics system and the electronic systems, a great deal of care must be exerted to guarantee that the subcontractor has experienced personnel for each of the different areas.

Other types of manual or publication are the operators and installation manuals, each specified to permit personnel to perform the specific tasks for which they were intended. The same comments made in conjunction with the maintenance manual would be applicable.

20.4 Spare Parts

For a procurement which involves development and for which the design details have not been established at the time of contract award, the identity, type, number, etc., of the spares to be furnished would be unknown. Therefore, that contract item would be left open as far as the specific list of spares and their cost is concerned.

After the design has been established and the parts and components to be used in the equipment have been determined, the selection of the list of spare parts for the equipment would be made as a coordinated effort between the contractor and the customer. Usually, the contractor would submit a recommended spare parts list as a deliverable item, and this list would be used for provision of the spares for the equipment.

The project engineer's main concern as far as the spare parts are concerned is that adequate numbers of the correct components be selected as spares so that the equipment operation and utilization in the field will not be jeopardized. If the equipment must remain inoperative due to the fact that some component was not included as a spare, the general opinion of the adequacy of the equipment or its "image" would suffer and would be a reflection on the manufacturer and ultimately on the project engineer.

In general, the guide to what part should be provisioned as a spare depends on its life expectancy. Here, the information derived from the reliability study would be used as a basis for selection.

Another factor to be considered is where the equipment will be used. If the equipment will be used at some isolated base, a comprehensive spares provision will be made. The area of utilization will be converted to a time factor. In an isolated base, the time factor might be 2,000 hours of average equipment operation. At a home base where sources of parts are accessible, a time factor of 1,000 hours might be used. Thus for a component having a meantime between failures of 500 hours, twice as many units of the component would be supplied for the isolated installation. Further, the isolated installation was to be provided with spares having expected lives of between 1,000 and 2,000 hours, whereas the home base installation would not be provided with such components as spares.

The RLS presents a particular problem as far as spares are concerned due to the optical system with its fragile lenses, mirrors, and other similar components. These items are not subject to wear, but spares should be provided due to the fragile nature of the items.

20.5 Drawings

The drawings constitute another item which involves a significant effort and a major cost factor. The method and organization necessary to implement the drawings were discussed in Chapter 15.

The only additional point to make is that the project engineer should make sure that the chief draftsman has analyzed the format and other requirements for the drawings so that there will be no rejections of the drawings as delivered on the contract due to the fact that the specification requirements were not followed.

20.6 Other Side Items

A contract can require the delivery of a multitude of different side items, all of which are important and require a serious effort by the contractor. Some other items which might be required on different contracts are installation services, maintenance services, and training of personnel in maintenance and operation of equipment.

It is important to remember that a timely and well-planned effort for providing the side items will minimize the difficulties and friction with the customer during the course of the program.

20.7 Summary

The deliverable items of a contract intended to serve as tools for the monitoring of the contract and to support the basic equipment in the field are referred to as side items. These items include manuals, spare parts, drawings, maintenance services, instructions for customer representatives regarding the operation and maintenance of the equipment, and other deliverable items.

The amount of money represented by the side items is usually about 10 percent of the total contract price which for a large procurement represents a significant amount. The contract and specification requirements for the side items are generally well-defined and call for strict conformance to details such as format, content, and delivery. Contractors in their zeal to design, develop, and fabricate the major item of the contract, which is the hardware, have been known to render only slight attention to the requirements of the side items. In many such cases, late delivery of the items and rejections due to failure to satisfy the contract requirements have caused a heavy penalty in unnecessary expenditure and damaged reputation because adequate attention and effort were not given to the requirements of the side items.

The successful delivery of the side items of a contract requires that the project engineer establish a course of action and schedule for all the items to guarantee that the specification requirements will be met and that each item will be delivered in accordance with the contract schedule.

Chapter Twenty-one

Follow-up

21.1 Postacceptance Considerations

The responsibilities of the project engineer do not cease when the equipment required by the contract is accepted and delivered. There are numerous areas of a contractual nature that require final settlement: residual items of the contract may still have to be delivered, and on a procurement of complex equipment the contractor may have a continuing function to maintain the equipment, provide changes desired by the customer, continue a Configuration Management Program for the delivered articles, and render other services desired by the customer.

21.2 Finalizing the Contract

Most contracts for equipment of a high cost incorporate some form of progress payment clause wherein the contractor receives payments of up to 85 or 90 percent of the cost of the equipment item being procured. The residual money is not automatically rendered to the contractor upon acceptance. Instead, the contractor and customer review the history of the procurement, the performance of the equipment, and other terms of the contract to determine what consideration should be offered as compensation in areas of noncompliance. Some of the types of issues in which the project engineer representing the customer and the contractor might become involved during the postacceptance negotiations include the following:

1. *Equipment performance:* The equipment may not have met every last requirement of the specification but may have been accepted because it was judged to be in "substantial compliance" with the contract requirements. In such a case, the customer, having received less than what was contracted for, is entitled to some consideration which would be equitable compensation.

2. *Latent defects:* A latent defect of the equipment is considered to be one that became apparent only after utilization but exists due to a flaw in the design or quality of the equipment. The contractor has the obligation to remedy the defect or negotiate equitable compensation to the customer.

3. *Delivery:* Delays in delivery may have been responsible for inconvenience or expense to the customer. If the delays were the contractor's fault, compensation would be negotiated.

4. *Claims:* Either party may have a basis for claims on the other due to some form of breach of contract. The customer, for instance, may have directed the contractor to provide some service or feature over and above the terms of the contract and therefore is entitled to compensation. Also, the customer may have failed to provide data or other items, as scheduled.

5. *Rate Structure:* Since most contracts for newly developed or engineered equipment are of the incentive type, customers, particularly the United States government, retain the right to negotiate overhead rates, G&A rates, and other factors that make up the final contract price.

The project engineer would participate in the final negotiation where each item is considered point by point and where a determination of the final price and possibly a revision to the schedule might be negotiated. The question might be asked as to the logic of negotiating a final schedule after the equipment has already been delivered. If the equipment is late, contractors would be interested in negotiating a schedule revision in order to have their performance record look favorable. A successful contractor must always look to the future, and one important negotiation point is the past performance by a company on previous contracts.

21.3 Subsidiary Items

Many of the subsidiary or side items require delivery subsequent to the primary equipment in the contract. The customer's project engineer has the responsibility for inspecting and verifying that the delivered item meets the contract requirements. Examples of such items include the following:

1. *Drawings and support documents:* One of the disciplines of Configuration Management discussed in Chapter 13 is to conduct a configuration audit to verify that the drawings and other documents are consistent with the equipment such data is supposed to describe. The audit, sponsored by the project engineer, establishes the acceptability of such material.

2. *Spare parts and support equipment:* When such items are required by the contract, the identity, quantity, and quality must be verified prior to rendering acceptance.

3. *Maintenance services:* The customer may require that the contractor render maintenance service on a continuous or on-call basis for a specific period of time. The project engineers for both parties usually are involved in implementing the maintenance program.

21.4   Field Reports

After the equipment is installed, the contractor will usually either be under contract to maintain the equipment or have the responsibility to correct difficulties that may occur during the warranty period which may last between 3 months and a year. In either case, the contractor would have access to the field reports describing the operation of the equipment.

The field reports are compiled periodically and describe the number of hours the equipment was utilized, difficulties experienced, corrective actions, and general comments. The project engineer would analyze each report and catalog the information to bring out any trends or patterns. For instance, if the reports indicate a recurring breakdown in a particular subsystem and the remedial action involved replacement of a particular component, an analysis of the design would probably reveal a need for a redesign of the subsystem which would be implemented as a field retrofit.

21.5   Unsolicited Proposals

An unsolicited proposal can be submitted at any time, and in the case of the United States government, there is an obligation to review the proposal and render an evaluation. Whether anything further materializes depends on the requirement that may exist and the funds that are available.

If a company has something to offer that seems to meet a need and is worthy of pursuing, the groundwork should be laid by communicating with the personnel of the potential customer and by promoting the merits of the equipment to be offered.

In the case of the RLS, the Acmen Electronics Company should submit any design improvements that may correct some inherent weakness as evidenced from the field service reports or which may improve the capability of the equipment. Such recommendations could be submitted as an unsolicited proposal. A close surveillance of the operation of the equipment and close liaison with the customer will serve to maintain good customer relations and enhance the chances of obtaining future contracts.

21.6 Summary

Project engineers have a continuing role after all items of the contract are delivered. The functions that would constitute the follow-up actions are settlements of claims, renegotiation, and other contractual issues.

Other follow-up actions would involve the receipt and correlation of field reports on the delivered equipment, the maintenance liaison with the customer, and a general effort to enhance their company's image in the eyes of the customer.

Glossary

Acceptance Tests: Tests to which the equipment is subjected to detemine whether the equipment meets the specification requirements.

Acquisition Phase: Period in Configuration Management Plan when item is being created either by in-house or by contractor effort. Phase is divided into design/development stage and production stage which terminates at operational baseline.

Activity: An element of work effort used in conjunction with PERT.

Allowable Costs: Costs incurred by a contractor for expenditures outside of the scope of the contract.

Armed Services Procurement Regulations (ASPR): A compilation of the rules that govern the procurement of equipment and services by the United States government from private industry.

Assembly Language: A computer code expressed in generalized computer language which is fed into the computer to achieve desired calculations. The computer itself will translate the assembly language into machine instructions.

Availability Factor: Measure of the ability of equipment to be utilized, expressed as functions of reliability and maintainability.

Baseline: Specific point during Configuration Management cycle which divides one phase from the following phase. Baselines, defined as characteristic, functional, and operational, terminate the concept formulation, definition, and acquisition phases, respectively.

Bit: A binary digit or a single character in a binary number. A word length of 6 bits such as - 100101 represents the number 37.

Block Diagram: A graphic presentation of a design in which the functions of subsystems or modules are shown as blocks, and interconnecting lines represent the flow of signals.

Boiler Plate Clauses: Standard contract clauses used in contracts by procuring organizations.

Breadboard: An energized interconnected assembly of components which is used to verify the design of a system.

Budget: Planned expenditures and commitments, by time periods.

Cash Flow: Money coming into and out of an organization or project as income and expenditures. Net cash flow is established after charges such as depreciation, interest, etc., are considered.

Changes Clause: Contract provisions permitting the customer (usually the government) to unilaterally revise the contract. Such changes, however, entitle the contractor to receive equitable adjustment in the contract price.

Characteristics Baseline: A point in the Configuration Management cycle which terminates the concept formulation phase and initiates the definition phase.

Clarification: Answers to questions posed by the offeror or the customer regarding ambiguities, incomplete material, etc., in specification proposals or other procurement material.

Commercial Considerations: Nontechnical factors which may influence a contract award such as price, cost sharing, arrangements, penalty/incentive issues, etc.

Commonality: A factor reflecting a percentage of components, modules, and other elements of a piece of equipment which are interchangeable with existing equipment of a different type.

Compatibility of Systems: The ability of two or more systems to perform effectively when interconnected electrically, mechanically, hydraulically, etc.

Compiler Language: The computer subsystem capability of accepting high-level computer language such as Fortran and effecting the translation into assembly language or into the machine language that can be used directly by the computer.

Computer Memory: A medium in which information relating to a computer can be stored on command and held for retrieval upon command, e.g., disc storage, core memory, etc.

Computer Program: The instructions which are used to direct the operation of the computer. The instructions can be expressed in compiler, assembly, or machine language.

Computer Speed: The rate at which a computer element can function. A computer speed of 1 microsecond indicates the cycle time of 1 million per second.

Concept Formulation Phase: The period in the Configuration Management cycle during which the desired objectives for satisfying a requirement are determined as feasible or unfeasible. The results of this phase establish the characteristics baseline.

Configuration Management: The implementation of formal management, technical direction, and controls during a project's life cycle to provide a complete definition of functional and physical characteristics of each item, to control the adoptions of changes, and to maintain a continuous accounting of the design and equipment configuration.

Constraint: The impediment in a PERT network caused by a particular activity which prevents other events from being reached or other activities from being initiated.

Constructive Changes: A communication by a representative of the procuring organization which is taken as direction by the contractor and acted upon.

Contingency Factors: An increment of cost or other margin to provide protection from unplanned expenditures brought about by unforeseen factors.

Contract Clauses: Terms relating to specific rights and obligations of the parties to a contract.

Contract Law: The legal ground rules that are applicable to parties in a contract which can be enforced or resolved in the judicial system of the government.

Contract Schedule: A contractual document specifying deliveries, description of deliverable items, terms of payment, special contractual requirements, and other items pertinent to the contract.

Contractor Claim: A request for consideration brought about by effort or expenditures for features or performance capabilities claimed as beyond the scope of the contract.

Cost Breakdown or Cost Segments: A segregation of total costs to identify the amounts required for specific efforts or expenditures.

Cost Effectivity: The total cost sustained by a user as the result of purchasing a particular type of equipment. The total includes those costs that might be imposed by modifications or changes to existing supporting equipment necessary to utilize the newly purchased items.

Cost Management: Procedures by which expenditures of resources and money are monitored and controlled.

Cost-plus-fixed-fee (CPFF): A type of contract generally used for R&D programs by which the contractor is reimbursed for all costs incurred, but the fee or profit is a fixed amount.

Cost Prediction Report: A report in which a comparison is made of the actual and budget costs at a point in time, and the trend of costs is indicated to facilitate cost prognostications.

Cost-reimbursable Contract: A type of contract in which costs sustained by the contractor are reimbursed. This type of contract normally is used for R&D tasks—the cost-plus-fixed-fee is a variation of this type of contract.

Cost Sharing: A formula used in incentive-type contracts in which a contractor derives a share in any amounts that are less than the target cost that the contract specifies. For any contract costs in excess of target, the contractor absorbs part of the excess costs.

Critical Path: The path of a PERT network which required the longest time to complete.

Cycle: A period during which the segments of events are accomplished leading to a concluding objective, e.g., Project Procurement Cycle, which is the time from contract award to acceptance of final contract item.

Definition Phase: The phase in the Configuration Management cycle during which the requirements previously established are translated in a specification and associated planning documents.

Detection System: A system designed to sense different signals and respond in specific ways in accordance with the signal characteristics.

Digital Computer: A machine comprising a multitude of on–off switches that are controlled to execute desired calculations.

Direct Costs: Specific expenses incurred on a project. Such expenses are engineering, manufacturing, and materials used on the project as contrasted to administration, depreciation, etc., that are indirect expenses.

Directivity Effects: As applied to radar returns, those effects caused by the relative position of the reflecting surface to the line of signal return to the antenna.

Display System: A system on which acquired information is presented in an intelligible form.

Drafting Order Revision: A directive issued to the chief draftsman by the project design engineer to reflect a change in the drawings.

Drafting Work Order: A directive issued to the chief draftsman to complete a drafting task.

Earliest Event Time: The earliest date that can be anticipated for the completion of specified work effort necessary to reach an event.

Effective Failure Rate: The failure rate of a component as modified by its application, conditions of operation, and environment.

Effectiveness: A measure of the probability that a piece of equipment will be ready and capable of performing its function.

Engineering Order (EO): A directive to the engineering department specifying an engineering task to be performed, its schedule, the budget of cost and hours, and other pertinent information to permit implementation of the engineering task.

Environmental Tests: A special series of tests whereby the equipment is subjected to different conditions of operation such as vibration, shock, and cold.

Evaluation: An analysis, using set criteria, to establish the relative excellence of documents such as a proposal.

Evaluation Factors: As applied to proposals to be evaluated, the factors represent the areas and criteria to establish the relative excellence of each proposal in each area.

Event: A specific point in a PERT network representing the start or completion of an activity. An event does not have any dimension in time or effort.

Excusable Delays: Failure to meet contract delivery dates due to circumstances beyond the control of the contractor, e.g., national catastrophies, failure of customer to meet contracted obligations, etc.

Expected Elapsed Time: The period of time that is predicted for completing an activity, generally used for PERT.

Fixed-price Incentive (FPI): A type of contract in which the target price is established with a sharing arrangement between the buyer and the seller for any amounts which the contract cost may fall under or over the contract target cost.

Flow Diagrams: A means of graphically displaying the mathematical model of a system including feedback, etc., in a simplified line diagram. Used as a design tool.

Flying Spot Scanner (FSS): An electronic tube in which the motion of the electron beam is controlled and directed to impinge on an object in precise patterns.

Fortran: A programmed computer language expressed by arithmetic formulae which requires translation by the computer into usable machine language.

Functional Baseline: The point in the Configuration Management cycle which terminates the definition phase and initiates the acquisition phase.

Furnished Property: Equipment, data, or other material that the customer is contractually obligated to provide to the contractor in a specific time frame.

Gantt Chart: A graphic presentation in which the data are shown as bars, and actual versus planned accomplishment is readily observed.

General and Administrative Rates (G&A): Rates applied to the direct and indirect overhead costs of a contract to cover all costs that cannot be identified with any specific program of a company.

Government Furnished Property: Items furnished by the government to be incorporated into the equipment to be delivered by the contractor or used in fulfilling the contractual requirements.

Grid Lines: A complex of lines generally vertical and horizontal which divide an area into small subareas.

High-level Program: A computer program expressed by standard arithmetic formulae such as Fortran, Cobol, etc.

Hour Breakdown: The hours of engineering, manufacturing, or other types of labor that would be applied to a project that is identified with specific tasks or types of effort.

Implementation Plan: A description of the resources and plans as to how an organization proposes to reach a particular objective such as fulfilling terms of a contract.

Infringement: The violation of a right of a party by another party, e.g., patent infringement.

In-house development: A project undertaken by an organization with their own resources.

Integrated Logistic Support: The general area of support for equipment used in the field. Support includes spares, instruction, manuals, maintenance, etc.

Integrated Tests: Tests to verify the performance of several interrelated systems when functioning together as a unit complex.

Iteration Rate: The speed at which a digital computer can perform all the required calculations for a result before repeating its cycle.

Latest Allowable Time: The latest date on which a PERT event can occur without delaying the completion of the program.

Latest Completion Date: The latest date on which a particular effort can be completed without effecting a delay in the program completion schedule.

Life Cycle: The period of time from the start of utilization of a piece of equipment until it ceases to have utility.

Line of Balance (LOB): Management tool for reporting the status of a project or subdivision of a project.

Loading: An amount of effort for which an organization is obligated during a specific period of time.

Machine Language: A language that does not require translation and that can be used directly in the computer.

Maintainability: The ability of a piece of equipment to be maintained or repaired. Maintainability can be expressed by the term, "Mean Time to Repair."

Make or Buy: Consideration of different factors to determine whether to purchase a part or system or undertake its design and fabrication with available resources.

Manpower Loading Display: A graphic presentation of personnel requirements by skill category for specific period integrals.

Material Breakdown: Identification of components, modules, and material with their costs.

Maximum Limit Conditions: Capacity of a component beyond which failure occurs.

Meantime between Failures: The elapsed time that a particular type of component in a system can be expected to function before failure. The elapsed time is obtained from the probability curve derived from the statistical distribution of previously observed or calculated data.

Memory Unit: A subsystem of a digital computer which functions to store a specified amount of data until required.

Need to Know: The justification of a person or an organization to obtain access to a particular area or type of information which is often classified.

Negotiations: Discussions between parties regarding the requirements of a proposed contract. The discussions are designed to arrive at a meeting of the minds so that the contract can be formally executed.

Network: A graphic presentation in which the tasks or functions which occur sequentially or in parallel are shown as an array of interconnected lines.

Objective Chart: Element of the Line of Balance Report which presents the percent effort of a task that is completed as of the report date.

Open: A term generally used in conjunction with electrical work which describes a condition such as a break in a line which results in an interruption in the flow of current.

Operational Baseline: The checkpoint in a Configuration Management cycle which establishes the equipment and its support item as ready for utilization.

Operational Phase: The Configuration Management period during which the equipment produced in a project is utilized.

Overhead Rates: A percentage added to a direct task cost, such as manufacturing labor, to cover indirect costs that are incurred in performing the particular task.

Packaging: The physical arrangement of elements of a system, its housing, ventilation, and other features of the hardware.

Pareto's Law of Distribution: A statistical theory which states that in any system, the greatest portion of the total cost is represented by a small percentage of the total number of subsystems.

Performance Specification: A specification which describes what is to be derived from the equipment but does not describe how the equipment is to be designed or fabricated.

PERT: Program Evaluation and Review Technique.

Phasing: Scheduling of different types of effort to promote an efficient flow of work, particularly for interdependent areas of effort.

Program: A means used in a digital computer by which the functions and operations of the computer are directed and controlled.

Progress Payments: A formula established in a contract which provides for payment to the contractor as various phases are completed.

Proprietary: As used in contractual sense, a sole source procurement.

Quality Control: The implementation of the procedures and means by which the quality of an element or system being produced is maintained and screened.

Radar Shadow: The shadow cast by an object as a result of illumination by a source of radar transmission. Any other object in the radar shadow will be obliterated in the resultant radar presentation.

Redetermination: A feature of a contract by which the customer reserves the right to review the costs and rates that are claimed by a contractor and to implement an adjustment in contract price as necessary.

Reliability: The ability of a component, system, or piece of equipment to function in accordance with a specified standard for a particular period of time.

Request for Proposal (RFP): An invitation extended to an organization to submit a proposal for a particular procurement.

Risk Factor: A percentage added to the estimate of hours or other costs of a project to cover unanticipated expenditures that might statistically be expected.

S Curve: A graphic presentation of some accomplishment as a function of time, the characteristics of which are S-shaped.

Schedule Prediction Report: A report in which the actual and estimated schedules are compared and the trends of progress are indicated to facilitate schedule prognostications.

Shadow Computer: A simulator computer which functions to generate a blanking pulse which is consistent with the geometry of the location and motion of the radar transmission, the location and dimensions of the object being illuminated, and other significant factors.

Short: A term used in electrical design relating to a condition between two points which offers no resistance to the current flow.

Side Items: The items of a contract, such as manuals, which are to be used to support the main item of procurement.

Simulator: A system which synthetically creates the functions and/or appearance of an operational system, thereby duplicating such equipment.

Slack: A measure of how much excess time is available or anticipated over and above the scheduled time for the completion of an activity or a series of sequential activities.

Slack, Negative: A measure of how much slippage exists or is anticipated in relation to the original schedule for an activity or a PERT path.

Slack Path: The measure of slack for a particular path of a PERT network.

Spoilage: Material spoiled in a manufacturing and fabrication effort that must be discarded as not usable.

Standard Deviation: A statistical measurement derived from the normal distribution curve which relates to boundary points for 68 percent of the total area under the distribution curve for one standard deviation.

Stress Ratio: Ratio of operating load over the rated load of a component.

Substantial Compliance: A legal term related to contracts which states that even when certain requirements are not achieved, the contractor cannot be held in default if the primary objectives are acceptable. The contractor would generally have to provide consideration to compensate for the deficiencies.

Subsystem: An integral portion of a large system that can be functionally separated.

Subsystem Tests: Functional tests on selected portions of a system that have to be isolated for purposes of the testing.

System Engineering: An overall design approach to assure that the various subsystems of a design will be compatible and result in the acceptable performance of the complete system.

Target Cost: The cost figure of a contract that represents the cost norm or goal.

Target Profit: The nominal profit that a contractor would realize if the contract was to be completed at the target cost figure.

Task Assignment Order: A directive issued by the project engineer to complete a specified mission within a planned schedule and budget.

Task Number: An internal number assigned by the contractor to a program which is used for cost accounting, purchasing, production, engineering, and all actions taken in conjunction with the contract.

Technical Proposal: A technical document presented by an organization describing how it is proposed that specific objectives will be accomplished.

Technical Proposal Requirements (TPR): A description of what type of information is to be presented in a technical proposal, the order of presentation, the format, and other requirements.

Termination: An action that brings all effort on a contract to a halt.

Test Procedure Report: Mutually accepted criteria to be used in the acceptance testing of the procured equipment.

Tiers of Design: Classifications of the degree of design detail of a particular equipment or system.

Time Function Conditions: Life span of a component under specific conditions which, when exceeded, results in fatigue failure.

Transparency: A medium which acts to modify light sources, thereby transmitting intelligence in the medium to some distant point.

Value Analysis: A program by which the savings realized as a result of recommendations made by a contractor are shared between the contractor and the customer.

Verification Test: A test to determine whether the equipment is capable of performing as required by the specification or test plan.

Word: A unit of coded information expressed as a series of bits which is used in a computer. A 24-bit computer indicates the computer is designed to represent the binary value expressed by a word length of up to 24 bits.

Work Package: The effort required to complete a specific task within an operating unit of the PERT system.

Index

Index

Acceptance tests (*see* Tests, acceptance)
Allowable costs, 87–88
Armed Service Procurement Regulations (ASPR), 56
Availability, 208, 219

Bits, 202–205
Block diagrams, 20, 21, 64, 185–187
Breadboards, 83, 190–191, 231, 232
Budget, responsibility for, 4, 5

Cash flow, 3, 5, 123
Changes clause, 86–87
Claims, 57, 59, 80
Commercial considerations, 8, 70–72
Commonality, 27–28
Communication(s), 168–181
  with the customer, 178–179
  of decisions, 171, 180, 181
  of instructions, 171–178
  project status, 170–171
  schedule and cost coordinator, 177–178
Communication cycle, 168–170

Compatibility, 26, 185
Components, 68, 213–217
Computer program, 200, 201, 231
Computer program tests, 233–234, 238
Computers, 12–15, 198–207, 231, 233–234, 238
  assembly language, 233
  bits, 202–205
  compiler, 233, 234
  Fortran, 234
  high-level program, 234
  iteration rate, 206
  machine language, 233
  memory unit, 14, 203
  speed of, 14, 200, 203–205, 234
Configuration Management, 146, 155–167
  baselines, 161, 165, 166
  benefits of, 165, 166
  changes, 164–166
  disciplines of, 162–164
  phases, 157–162
  requirements for, 158
Constructive changes, 121–122
Contingency factor, 62, 63, 70, 71, 75, 82, 83, 108

Contract clauses, 86–98, 109, 110, 122–124,
    194, 195
  allowable costs, 87–88
  boiler plate, 86
  changes, 86–87
  default, 96
  defects, correction of, 88–89
  delays, excusable, 91, 96
  disputes, 91–92
  Government Furnished Property (GFP),
    92, 96
  liquidated damages, 58, 124
  overtime, 93–94
  patents, 92–93, 97
  penalty, 109–110, 124
  subcontracts, 89
  termination, 89–90, 122–123
  value engineering, 94–95, 194–195
Contract documents, precedence of, 51, 53,
    79, 86, 120–121, 146
Contract law, 3, 114–124
Contract schedule, 51–59
  analysis of, 54–59
Contracts, types of, 74–85
    cost reimbursable, 74, 79–84
      cost-plus-fixed-fee (CPFF), 75, 79–
        81
      cost-plus-incentive-fee (CPIF), 75,
        79–81
    fixed-price, 74–79, 82–84
      with escalation, 76
      incentive, 77–78
      redetermination, 75–77
      two-step, 78–79
Copyright, 92–93
Cost(s), 6, 22, 23, 62–73, 77, 78, 82, 83, 99–
    113
  allowable, 87–88
  breakdown of, 6, 44, 63
  ceiling, 77, 78, 109, 110
  comparison of, 26, 27
  competitive, 22, 23, 70–72, 82
  contingency, 83, 108
  direct, 65–69
  engineering, 64–67, 83
  estimating of, 60–73
    pitfalls of, 62, 63
  manufacturing, 68–69, 83
  material, 67–68, 83
  negotiation of, 99–113
  overhead rates, 69–70, 108
  sharing, 77, 78
  target, 77, 78

Decision making, 179, 180
Default, 96
Defense Documentation Center, 65
Deficient compliance of specifications, 118
Delays, excusable, by contractor, 91, 96
Delivery schedules, 2–8, 28–29, 33, 43–47,
    52, 55–59, 91, 101–105, 125–143, 148,
    149, 177, 178
Design, 11–29, 83, 182–195, 241, 242
  compatibility of, 26, 185
  data for, 7, 183, 184
  packaging, 191–192
  planning, 19–26, 147–149, 182–194
  for standardization, 187–189
  system, 184–187
  of the subsystem, 187
Design approaches, 12–16, 18, 24, 26
Design approval, 53, 57
Design detail, tiers of, 19–21
Design freeze, 241, 242
Disputes clause, 91–92
Drafting, 192–194, 224
Drafting Order, 175
Drawings, 192–194, 240, 244
  engineering, 192, 240
  manufacturing, 192

Effectiveness, 208, 219, 220
Elapsed time, 130–135
End Items, Contract, 160
Engineering Change Form (EC), 173
Engineering Order (EO), 173
Environmental tests, 231, 236–237
Escalation type of contract, 76
Evaluations, 34, 35, 41, 42
Excusable delays, 91, 96

Facilities, 7, 33, 36, 41, 46–48
Failure rate, 209, 213–218
Field reports, 248
Flying Spot Scanner (FSS), 21, 24–26
Fortran, 234
Furnished property, 53, 56, 57, 92, 96, 123

Gantt Chart, 49, 55
General and Administrative (G&A) rates, 7,
    42, 43, 69, 70
Government Furnished Property (GFP) (*see*
    Furnished property)

High-level computer language program, 234

Implementation plan, 31
Inconsistencies in specifications, 117
Infringements, 92, 93, 97
In-house task, 150–152, 154
Instructions, 171–177
 drafting, 174–175
 engineering, 172–173
 production. 176–177
 purchasing, 175–176
 quality-control, 173–174

Latent defects, 247
Legal considerations, 114–124
 appeal procedures, 114–116
 contract document, 120–123
 evaluations, 118–120
 specification deficiencies, 118
Life cycle of contract, 75
Line of Balance (LOB), 139–142
Line items, 52, 54
Liquidated damages, 54, 59, 124
Logistic support, 31, 239–244

Maintainability, 218–221, 231
Make or buy decisions, 4, 150–152
Manuals, maintenance, 242–243
Mean Time Between Failures (MTBF), 10,
 209, 216, 217, 219
Mean Time to Repair (MTR), 209, 218
Monitoring items (see Production,
 monitoring)

Need-to-know, 65
Negotiation parameters, establishing, 101–
 105
Negotiation tactics, 110–112
Negotiations, 99–113
 analysis of, 101–106
 over clauses, 108–110
 costs of, 101–105, 107–108
 definition of, 100–101
 objectives of, 99–100
Negotiator, qualities of a, 112–113
Network, PERT, 129–133
Nonresponsive bidder, 100

Obsolescence, effect of, on equipment,
 210
Omissions in specifications, 117
Organization, project, 144–148
Overhead rates, 42, 43, 69–70, 151
Overtime, 93–94

Pareto's Law of Distribution, 62
Patents, 92, 93, 97, 119
Penalty clause, 109, 110
PERT (see Program Evaluation Review
 Technique)
Phasing, 28, 29, 45, 55, 56, 224
Plant load, 46–48
Precedence of documents, 51, 58
Probability features of PERT, 133–136
Production, 42–47, 68, 69, 139–143, 150,
 165, 176, 177, 222–227
 monitoring, 139–143, 149, 150, 165, 224–
  227, 239–240
 quality control in, 227–228
Production instructions, 176, 177
Production planning, 223–227
 schedule for, 45, 46, 55, 223–227
Production Work Order, 176, 177
Program Evaluation Review Technique
 (PERT), 125–143
 constraint, 129–131
 cost control, 136, 137
 critical paths in, 129–133
 cycle of, 127–129
 definitions of, 126, 127
 establishment of, 129–131
 manpower loading, 137, 138
 network, 129–133
  use of, 131–133
 probability features of, 133–136
 reports, 137–139
 slack paths in, 126, 131
 variance, 135, 136
Program status, 137–143
Progress payments, 123, 124
Project engineer:
 authority of, 4
 objectives of, 2
 responsibility of, 4
Project organization, 145–147
Project phasing chart, 148, 149
Project Production Order (PPO),
 176
Project tasks, 147–148, 152–154
 scheduling of, 148–149

Proposal(s), 31–50
    costs of, 42–44
    evaluation of, 41, 42, 118–120
    introduction of, 37
    outline of, 35, 36
    Request for (RFP), 6–7, 31–35
    unsolicited, 248–249
Proposal language, 36–37
Proprietary rights on government
        procurements, 119
Prototype equipment, production of, 222–
        223

Quality control, 227–229
Quality-control instructions, 173–174
Quality Control Order (QCO), 173

Rates, 3, 69, 70, 150, 151, 247
    G&A (*see* General and Administrative
        rates)
Redetermination, 76–77, 84
Reliability, 189–190, 208–221, 231
    components for, selection of, 213–214
    in design, 211–213, 217
    effectiveness of, 219, 220
    failure rate for, 213–217
    prediction of, 216, 217
    stress factor for, 216–218
    testing, 212, 213, 215
Reports:
    engineering, 52, 53, 56, 240–243
    field, 248
    management, 137–143
Request for Proposal (RFP), 6–7, 31–35
Risk, 6, 8, 60, 62–64, 71, 75, 101, 103, 108
    analysis of, 81–84

Scheduling, 3, 43, 45, 55, 128, 148–149,
        237
Shadow Computer, 63–66, 129–133, 177,
        211–217

Shakedown period, 210
Side items, 107, 239–244, 247
Slack paths in PERT network, 126, 131
Spare parts, 243–244
Specification(s), 6, 9–11, 19, 21, 22, 25, 26,
        117–118, 144, 145
    analysis of, 11–16, 25, 144, 145
Spoilage curve, 68
Standard deviation (SD), 134–136
Standardization, 187–189
Stress ratio, 216
Subsystem tests, 231, 239
System approach, factors affecting, 21–24
Systems, 18, 231

Target cost, 77, 78
Task Assignment Form, 152, 153
Tasks, 152–154, 182
Technical Proposal Requirements (TPR), 6,
        31–34
Termination clause, 89–90, 122–123
Tests, 33, 36, 41, 53–55, 87, 130, 194, 212,
        230–237
    acceptance, 53, 54, 130, 231–232
    breadboarding, 190, 191, 231–233
    computer program, 233–234, 238
    criteria for, 231–232
    environmental, 231, 236–237
    integrated, 231, 235, 236
    reliability, 189, 190, 236, 237
    scheduling of, 55, 237
    for subsystems, 190, 191, 231–233
Tiers of design detail, 19–21

Value engineering, 94–95, 194–195

Work Breakdown Structure, 19–22, 62–64,
        127, 129
Work package, 127, 223